Basic Biology

Charles LaRue

American Guidance Service, Inc.
Circle Pines, Minnesota 55014-1796

THE AUTHOR

Charles J. LaRue holds a Ph.D. in science education, zoology, and botany from the University of Maryland, and an M.A. in zoology and botany from the University of Texas. He has spent fourteen years as a biology teacher, two years as a zoology and botany instructor at the college level, and fifteen years as a supervisor of elementary science in the Montgomery County (Maryland) Public Schools. He is the author of *Basic Health*. He participated in the early development of BSCS Green Version of *Biological Science: An Ecological Approach*. He has co-authored fifty-five learning packages in science. He has also co-authored with Paul Brandewein *100 Investigations in Laboratory Science*. He serves as a consultant for the National Geographic Society for educational filmstrips, films and books.

CONTRIBUTORS

Elissa D. Weinroth, Ph. D. made major contributions to the chapters on human health and on genetics. She served as consultant and preliminary editor for the text.

Alan Weinberg, M.Ed., a special education teacher, contributed to the chapters on the lives of plants and the systems of the human body.

Stephen McDaniel and Mary Angela McDaniel, former biology teachers, and now professional photographers, contributed to the chapters on human reproduction and on ecology.

STAFF

Bookmark Associates, Editorial Services

Go Media, Page Composition

Norm Myers, Design

Printed in the United States of America.

ISBN: 0-88671-543-1 (Previously ISBN: 7916-0097-1)

Order Number: 80190

A 0 9 8 7 6 5 4 3

Contents

Preface vii

1 **Describing How Living Things are Alike** 2
  Cells: Basic Units
  Basic Life Chemicals
  Basic Life Activities

2 **Organizing All Natural Things** 14
  Living or Non-Living?
  Five Major Kingdoms

3 **Classifying and Organizing Animal Groups** 24
  Using Similar Structures to Classify
  All Animals Fit Somewhere in the System
  Animal With Backbones
  Animals Without Backbones

4 **Classifying and Organizing Plant Groups** 40
  Classifying Plants
  The Seed Plants
  The Ferns
  The Mosses

5 **Organizing Protists, Fungi, and Monerans** 52
  The Protist Kingdom
  Basic Life Activities of Protists
  Monerans
  The Fungi Kingsom
  Fungus Life Activities

6 Discovering How Animals Stay Alive 76

Food Getting and Ingesting

Digestion and Excretion

Circulation and Respiration

Coordination

7 Discovering How Plants Live Their Lives 98

The Transport System in Vascular Plants

Food Making in Green Plants

Plants Give Off Oxygen

Reproducing New Plants

8 Understanding Systems of the Human Body 116

Ingestion and Digestion

Body Systems and Life Activities

Coordinating Systems of the Body

The Human Skeleton

**9 Describing Patterns in Reproduction,
Growth, and Development** 146

Life From Other Life

Sexual Reproduction

Growth and Development of Animals

Human Reproduction, Growth, and Development

10 Understanding the Human Fight to Stay Healthy 174

The Human Fight Against Disease

Nutrition and Diet

11 Following Genetics From Generation to Generation 186

What is Heredity?

What Chromosomes Do

Human Genetics

Applied Genetics

12 Tracing Cycles in the Environment: Ecology 210

Ecosystems

Food Chains and Food Webs

Spaceship Earth

What Is Energy?

13 Describing the Behavior of Organisms 234

Species Behavior

Learned Behavior

Communication Behavior

14 Tracking Changes in Organisms Through Time: Evolution 254

Fossil Evidence

Evolution Is a Process of Change

Fossil Humans

Glossary 279

Index 299

Preface

What Is Biology?

Biology is a science that asks questions about what it means to be alive and about how living things do some of the things they do. Biology is about the relationships among all living things, and about their environments. Biology is about the insides of animals and plants — about how they are put together and how they work. Biology is about how animals and plants have changed over long periods of time.

If students ask why they are going to study biology, they are off to a good start. Being curious is a common human trait. Humans have been asking and wondering about animals and plants for a long time. People in general have a natural curiosity about themselves and other living things.

The Purpose of This Book

Basic Biology is a text book designed for those students who need a direct simplified presentation of the important concepts of biology. This book demonstrates the value of using the scientific system for classifying organisms. Readers will become more aware that the world is full of patterns. New information is discussed in such a way to show a connection with previous understanding. As the material unfolds, students move from the known to the unknown. They are encouraged to view their knowledge of basic life activities as being useful to them not only for the moment, but for tomorrow and the next day.

Organization of the Book

The main thread of the text is classification, animal biology, human biology, ecology, genetics, human reproduction, and evolution by natural selection. The fourteen chapters are each divided into meaningful key ideas. Each Key Idea section ends with questions to check understanding.

Chapter Conclusions

At the conclusion of each chapter there is a summary and a list of important words used in the chapter. The final exercise of the chapter is an extensive set of review questions. By answering these questions, students can determine if they have mastered the main ideas in the chapter.

Vocabulary

Concepts are presented in clear and simple language. Features that make the book easy to read are: large type, illustrations, generous headings, short sentences, and brief paragraphs. An effort was made to use familiar words and a comfortable style, while retaining essential biological terms.

Teacher's Guide

The Teacher's Guide that accompanies this text contains objectives, teaching suggestions, and complete answers. There is a substantial section of reproducible exercises and answer keys to all components.

Manual of Laboratory and Language Activities

The accompanying student *Manual of Laboratory and Language Activities* provides additional opportunities for skill development in manipulation, observation, measurement, description, and classification. Reading and vocabulary exercises promote fuller comprehension of biological concepts.

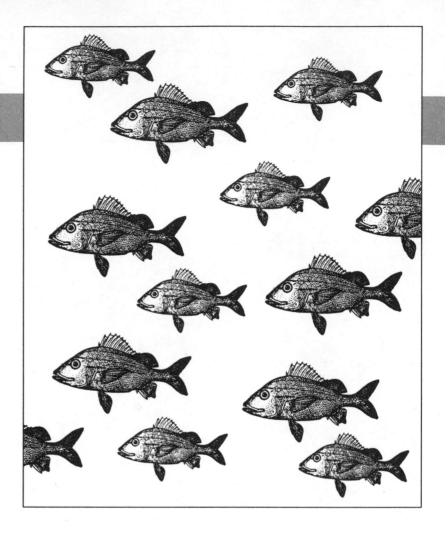

CHAPTER GOAL:

To describe three ways in which living things are alike: in their cells, in their chemicals, and in their life activities.

KEY IDEAS:

Living things are alike because they are made up of basic kinds of building blocks called cells.

Living things are alike because they have the same basic chemicals.

Living things are alike because they carry out the same kinds of basic life activities.

CHAPTER 1

Describing How Living Things Are Alike

Introduction

Cells, the Basic Units

Livings things are alike. Plants, animals, protists, fungi, and monerans are living things, and they are alike in several ways. They are alike in one way because they all have cells. Cells are the main parts of organisms. Cells are called **basic units**. Cells, or basic units, are used for building body parts and for carrying out life activities.

Chemicals

Cells are living material made up of chemicals. The same **basic chemicals** make up the bodies of all organisms. Living things use the chemicals to stay alive. They need to have enough of the chemicals, and they need to have them in the right place at the right time. Most living things use a chemical form of sugar to stay alive.

Life Activities

Living things are also alike in how they stay alive. They all carry out similar **basic life activities**. Getting food, eating, breathing, moving, digesting food, growing, reproducing, and sensing things are among the basic life activities of animals, plants, and other forms of life.

Organisms are alike in these ways:

- They are made up of basic units called cells.
- They are made up of similar chemicals.
- They carry out the same kinds of life activities.

KEY IDEA #1

Living things are alike because they are made of basic kinds of building blocks called cells.

Cells: Basic Units

Examples of Human Cells

Nerve cell

Muscle cells

Gland cells

What are cells? **Cells** are the basic structure of all organisms. Cells are the smallest units that can carry on the activities of life. They are invisible to the eye. Cells are the living parts of all life forms.

Size and Shape of Cells

Cells come in different sizes, from small to very small. They come in different shapes, such as triangles and squares. Cells also may be shaped like rectangles or circles. Cells may be long, short, wide or narrow.

Cell Functions

Cells carry out several **cell functions**, or jobs:

1. They cover (skin cells).
2. They help in movement (muscle cells).
3. They protect (bone cells).
4. They send and receive signals (nerve cells).
5. They transport or carry (blood cells).

Cells in Living Things

Cells are found in all parts of animals: in blood, bone, and muscle. Cells are in all parts of plants: roots, stems, leaves, and flowers. Cells are basic units in all organisms.

Cells Make Tissues

Cells are used for building the different tissues of organisms, such as muscle tissue, nerve tissue, and bone tissue. In animals there are muscle cells, blood cells, skin cells, nerve cells, and bone cells. Muscle cells are joined together in special ways to make different kinds of muscle tissue: leg muscles, arm muscles, stomach muscles, back muscles, hand and foot muscles, and heart muscles. Plants also have different kinds of cells, such as root cells, stem cells, and leaf cells. Inside a leaf there are different kinds of cells.

Groups of cells that are similar and act together to do some jobs are called **tissues**. Muscle cells work together to cause movement. The work is done by muscle tissue. Other tissues in living things are nerve tissue, bone tissue, and skin tissue. Living things are alike in the way they are put together with groups of cells called tissues.

Tissues Make Organs

Different kinds of tissues join together to form organs. Tissues are the main parts of **organs**. Organs are the main working parts of animals and plants. Organs do special jobs. Your heart is an organ. Your lungs are organs. So are your stomach, liver, kidneys, eyes, ears, bladder and your skin—the largest organ. Living things are alike because they have cells that form tissues and the tissues form organs. Organs are the main parts of living things.

Onion Skin Cells

Cell walls have cellulose.

Each cell has a nucleus.

Sense Organs

Eye

Ear

Skin

Plant Organs

Leaves

Stem

Roots

CHECK YOUR UNDERSTANDING
1. What are cells?
2. What are three functions of cells?
3. What are tissues made of?
4. What are organs made of?
5. Name four kinds of organs.

KEY IDEA #2
Living things are alike because they have the same basic chemicals.

Basic Life Chemicals

Besides having cells, living things are alike because they have similar chemicals. Living things use these chemicals to stay alive.

Water, Water, Everywhere

A Plant Cell

Cell wall
Cytoplasm
Nucleus
Cell membrane
Vacuole

Water is one of the chemicals used by all living things. Water is the most plentiful chemical in the bodies of animals and plants. Water is found in each of the trillion cells in the human body. For every one hundred pounds of your body weight, about sixty-eight pounds of that weight is water.

Water—The Great Dissolver

Water is both useful and necessary. Life cannot exist without water. Water **dissolves** other chemicals. To dissolve means to change something from big pieces to smaller pieces. When something dissolves, it gets broken apart. Water breaks things apart. Cells are so small that the materials that go in and out of them must be dissolved. Water is the chemical that does the dissolving. When material is dissolved into small and invisible pieces, it can move more easily from one cell to another cell.

Salt Water in Cells

The water in your cells is not pure or fresh water. It is a weak solution of salt water. The water in all animals is a mix of salts and water. Even fish found in fresh water streams and ponds have water with salt in their bodies.

Carbohydrates, Fats, and Proteins

Other important chemicals found in animals are carbohydrates, fats, and proteins. Each of these common chemicals has a job to do in the body of a living thing. **Carbohydrates** contain carbon, hydrogen, and oxygen. Look again at the word *carbohydrates*. The first part, *carbo*, stands for carbon, and the second part, *hydrates*, stands for hydrogen and oxygen. Carbohydrates are sugars and starches. They are found in foods such as bread, cakes, and potatoes. They are used by you and all other living things as a source of energy.

Chemicals for Energy

Energy is the power needed to carry on life activities. Energy comes from fuel. Carbohydrates are **fuel chemicals**. Carbohydrates in your body work like gasoline in a car. Gasoline from the fuel tank gets to the engine, where it is broken apart and energy is released. This released energy runs the engine. When the carbohydrates break apart in your body, energy is released. This energy powers your body. The same thing happens in other animals. Plants use energy, too.

Fats for Fuel

Fats are important chemicals found in foods such as beef, butter, cheese, and peanut butter. Fats provide the body with a source of energy. Fats supply a source of energy because they are fuel chemicals.

Proteins for Life

Proteins are another kind of chemical found in meat, chicken, fish, and beans. Proteins are part of all living things. Proteins do two main things:

- They help to build and repair body parts.
- They help control body activities such as heart rate, and the breaking down of food in the stomach and small intestine.

Nutrients and Your Diet

Keeping your body going is not a simple job. You must get a regular supply of carbohydrates, proteins, and fats from your food. It is important to get a *variety* of food in your diet. Each kind of food that you eat provides different chemicals—chemicals for energy and chemicals for growth. In addition to carbohydrates, proteins, and fats, your body also need minerals, vitamins, and water. These chemicals—carbohydrates, proteins, fats, minerals, vitamins and water—are all called **nutrients**. Living things need a fresh supply of these nutrients every day.

Living Things Are Alike

Can you see how living things are alike? Living things are made up of cells. Living things are made up of the same chemicals. Living things need and use the same chemicals. Living things need nutrients.

Some Sources of Nutrients

Water

Bread is one source of carbohydrates.

Meat is one source of proteins.

Butter is one source of fat.

CHECK YOUR UNDERSTANDING

1. Why do living things need water?
2. What are three important chemicals in animal bodies?
3. Of what special use are carbohydrates in your body?
4. What do proteins do in your body?
5. List the six nutrients that are chemicals of life.

KEY IDEA #3:
Living things are alike because they carry out the same kinds of basic life activities.

Basic Biology

Basic Life Activities

All living things carry on the same kinds of activities. Getting and digesting food, sensing things, getting oxygen, moving, growing, and reproducing are some of the basic life activities of organisms.

Getting Food

A common example of a basic life activity is getting food. Animals must get their food by eating plants or by capturing other animals. Bees and elephants, humans and whales, birds and sharks, all must get their food. Plants make their own food. Roses and maple trees, grass and corn, grapevines and orange trees, all make their own food.

Some animals eat other animals for food.

Green plants can make their own food.

Using Food and Removing Wastes

Digestion is a life activity that changes the food you eat into a useful form for cells to use. **Respiration** is another basic life activity. It causes oxygen to meet with the chemicals in food to release energy for body cells to do work. Respiration also gets rid of some cell waste matter. **Excretion** is the basic life activity that involves getting rid of solid and liquid waste.

Movement

Movement is another life activity that is common to all plants and animals. Plants do not move from one place to another, but they still move. Animal movement is easier to see. Plants have roots that hold them in one place, but their parts bend or move toward light. Besides outward movement, there is constant movement inside living things. The insides of plants and animals are always changing. Movement goes on and on. Blood is circulating, food is digesting, and breathing is taking place. Movement is one of the basic life activities.

Growing

Growing is one of the most obvious basic life activities. You have grown. You may have been a small or a large baby, but you have now grown to a larger person. You are still growing. You will continue to grow until you reach your adult size. Most living things go through a similar pattern. Growing is a part of being alive. It is one of the basic life activities.

Sensing and Reacting

Sensing and reacting are basic life activities. All living things behave in certain ways. Animals and plants

A sunflower turns toward the sun.

A moth flies to the light at night.

have tissues and organs that pick up signals from their surroundings. Plants and animals **react** or respond to these signals. The signals may be light, sound, smells, or touch. Some moths fly to lights at night, fish swim to the top of a tank for food, and your dog comes to you at the sound of your voice. Many flowers open in the morning light and close as the sun goes down. Sensing and reacting are things that all living things do.

Development

Development is another basic life activity. For example, a puppy goes through changes that are a part of growing up. It will develop a different shaped snout, a very special tail, and an entirely new voice box (from a whiner to a barker). **Developing** means getting bigger, but it also means becoming different. Tadpoles develop by stages into frogs and toads. They get bigger and at the same time grow different parts. Development is another basic life activity that happens to all living things.

Reproduction

Reproduction is a basic life activity through which organisms produce offspring, or children. However, reproduction is not always something that individuals can do alone. Most reproduction involves two parents. All living things have reproduction as one of their basic life activities.

CHECK YOUR UNDERSTANDING

1. What is the difference in the way that animals and plants get food?
2. What is the main difference between the way animals and plants move?
3. How long does a person continue to grow?
4. What is meant by "sensing things and reacting?"
5. Name three life activities that have to do with using food and removing wastes.

Summary of Chapter 1

Living things are alike in three ways. They have basic units called cells; they have sets of similar chemicals; and they carry out the same kinds of basic life activities.

Cells come in many sizes and shapes. Cells do a great deal of work in the body. They cover, help movement, line organs, send signals, and transport materials. Cells are mostly water. Water is the great dissolver of chemicals. Cells are the building blocks of tissues and organs.

The chemicals of life are water, carbohydrates, fats, and proteins. These chemicals serve as energy, and as building and repairing materials.

All plants and animals carry out basic life activities. These include getting and using food, moving, growing, sensing, reacting, removing waste, developing, and reproducing.

1. *cells*
2. *cell functions*
3. *tissues*
4. *basic units*
5. *organs*
6. *basic chemicals*
7. *proteins*
8. *carbohydrates*
9. *fuel chemicals*
10. *dissolves*
11. *fats*
12. *nutrients*
13. *sensing*
14. *respiration*
15. *reproduction*
16. *movement*
17. *excretion*
18. *basic life activities*
19. *digestion*
20. *developing*

REVIEW QUESTIONS FOR CHAPTER 1

1. Name three ways in which living things are alike.
2. What are three things that you can tell about cells?
3. What are the main functions that cells carry out?
4. Name three kinds of animal tissue.
5. What kinds of cells are found in plants?
6. What is the definition of the term "tissue?"
7. How are organs related to tissues?
8. What do fats do for animals?
9. What are the main things that proteins do for animals?
10. Name the basic life activities.

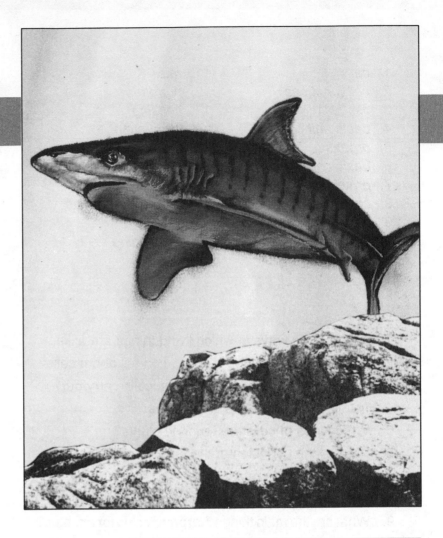

To explain the basic differences between living and non-living things.

To explain why the living world is divided into five separate groups called kingdoms.

KEY IDEAS:

Living things differ from non-living things in their characteristics and properties.

The living world is divided into five major groups: plants, animals, protists, fungi and monerans.

CHAPTER 2

Organizing All Natural Things

Introduction

Living and Non-Living

What is biology? **Biology** is the study of living things or organisms. What is an **organism**? An organism is a complete, individual living thing that must breathe, move, grow, react, reproduce, digest, and get and use food. An organism must carry out these basic life activities. Basic life activities make living things more complex than non-living things.

The non-living world is made up of things that are solids, liquids, and gases. Mountains, rivers, oceans, and air are all part of the non-living world. So are rocks, sand, and soil. Chemicals are also part of the non-living world.

Plants, Animals, Protists, Fungi and Monerans

Scientists have classified all the organisms in the world into five large groups: plants, animals, protists, fungi, and monerans. Most of the living things that you know are either plant or animal. But there are three groups that you probably do not know as well. These are the protists, fungi, and monerans. Protists, fungi, and monerans are separate from the plant or animal groups. But they belong to the living world, because they carry out the basic life activities.

Living Or Non-Living?

The entire world of **natural things** may be taken apart and put into just two groups—**living things** and **non-living things**. Living things are either plant, animal, protist, fungi, or monerans; non-living things are rocks, soil, water, or gases like oxygen and carbon dioxide.

Non-Living Things

A non-living thing does not carry out the basic life activities. The non-living things in the world are all of those basic materials that we call solids, liquids, and gases. Sand, water, and air are non-living. So are rocks, minerals, and vitamins. None of these things have any of the basic life activities.

Properties

In order to describe non-living things, we tell about their **properties**. Properties of non-living things are what you can see or find out about an object. The color of a rock is a property of that rock. How a rock breaks into pieces is another property. Each rock or mineral will scratch some things and not others because of its property of hardness. A scale of hardness can be made to identify each kind of rock or mineral. Did you know that a diamond is the hardest mineral of all? It is used in industry as the main part of some cutting tools because of that property.

Living Things

The living world is made up organisms. To qualify as a living thing, an organism must have **basic life activities** such as getting and using food, moving, and growing. Plants, animals and other organisms carry out the basic life activities. Plants, animals, fungi, protists and monerans make up what we call the living world. Plants and animals are the most familiar organisms. Fungi, protists, and monerans make up the rest of the living world.

The first part of the word *organism* is **organ**. Plants and animals have organs, or special parts, that are used to carry out the basic life activities. Simpler organisms like fungi, protists, and monerans do not require advanced systems of organs. Non-living things do not have organs.

Basic Differences Among Living Things

Living things are more complex or complicated than non-living things. There are also differences between different living things.

No two people are exactly alike, not even identical twins. A characteristic of living things is that each one is different from every other one.

Think about some living things that seem to be very much alike. In what ways are these things alike? In what ways are they different? Now think of some non-living things in the natural world that seen to be nearly identical. Pieces of coal, sand grains, and snowflakes may have come to your mind. There are fewer differences between grains of sand and snowflakes than between the puppies in a litter.

The idea the "no two things are exactly alike" is easier to see when you think about living things. The basic life activities of living things make living things more complex or complicated than non-living things. Finding differences between living and non-living things is one way to divide the whole world into two parts.

CHECK YOUR UNDERSTANDING

1. Into what two groups can everything in the natural world be divided?
2. What must something have or do in the world to qualify as a living thing?
3. What is an organism?
4. Name two differences between living and non-living things.

KEY IDEA #3

The living world is divided into five major groups or kingdoms: plants, animals, protists, fungi, and monerans.

Five Major Kingdoms

At one time, scientists divided the world of living things into just two main groups or kingdoms—the plant kingdom and the animal kingdom. Later, when biologists used the microscope, they discovered tiny organisms that did not fit into either the plant or the animal kingdom. A new kingdom was formed in order to have a place for these microorganisms. This kingdom was named protists. Later, another kingdom called fungi was added. In this system of classification, there were four main kingdoms in the living world. More recently biologists have used a five kingdom classification system. They have added a group called **monerans**.

The Plant and Animal Kingdoms

The two groups with which you are most familiar are **plants** and **animals**. What are some ways to tell the difference between between an animal and a plant? The squirrel and the dandelion are both living things, but they are different. The squirrel can move about. The dandelion stays in one place. The squirrel has to find its food. The dandelion makes its own food.

Getting Food

Animals and plants differ in the way they get their food. Animals get their food from other living things. They eat plants, or they eat other animals. Animals move around to capture or gather their food.

Plants make their own food from the materials they take in through their roots. They also use the energy they get from the sun. Trees, shrubs, grasses, and vines are easy to recognize as plants. They stay in one place and make their own food.

Protists, Fungi, and Monerans

The world of living things has three other groups of organisms that you probably do not know as well as plants and animals. These are protists, fungi and monerans. They are in separate kingdoms because they have their own characteristics. They do not fit into either a plant or an animal group. None of these resembles the organisms that you call plants and animals. But they do belong to the living world, because they carry out the basic life activities.

Animals and plants are quite different.

Bacteria

Paramecium

Euglena

Ameba

What Protists Are Like

Most protists are microorganisms. That is, they are very small and can only be seen under the microscope. The most common of the protists are the bacteria and the protozoans. Protozoans include parameciums, euglenas, and amebas. There are thousands of kinds of protozoans. Some protozoans have both plant and animal traits. They act like animals by moving about, but they also make their own food like plants.

Algae are protists that vary in size. They go from microscopic size to meters in length. Algae live in lakes, streams, rivers, ponds, and oceans. Algae of the oceans are the great food producers of the world.

The Importance of Protists

Protists are important to humans and other organisms. Protists include protozoa and algae. There are thousands of kinds of protozoa and algae. The algae of the living world are found in water. They produce food in great quantities for the organisms found in the waters of the world. In addition, algae produce most of the oxygen in the atmosphere. They do this because they are able to carry out photosynthesis. Photosynthesis provides food and at the same time gives off oxygen to the atmosphere. Green algae provide much of the oxygen in the world.

Protozoa are microscopic protists. Different kinds of protozoa have different methods of moving about in their water environment. They have flagella or "tails," cilia or "oars," pseudopods or "false feet." One group of protozoa does not move in the water like the others move. They cause disease in humans and are spread by the bite of a mosquito.

The Kingdom of Fungi

The fourth kingdom in the five kingdom classification system is the kingdom of fungi. (Fungi is the plural of fungus.) At one time fungi were classified with plants.

Basic Biology

However, they do not carry out one of the basic life activities of plants: they do not make their own food. Instead they take in, or **absorb**, food from the remains of other organisms. Neither do fungi carry out one of the basic life activities of animals and protists. They do not move around by themselves; they stay in one place.

Many fungi are **parasites**. Parasites are organisms that live on other organisms and depend upon them for their food. Sóme common fungi are mushrooms and molds.

The Great Decomposers

The most important job of fungi is to decompose or break down materials around them, so that they can be used by other organisms. Fungi decompose waste matter such as dead animals and plants and change them into usable chemicals. They return these chemicals to the soil where they can be used by other organisms.

The Kingdom of Monerans

Monerans are simple, one-celled organisms. However, they carry out the basic life activities. Monerans are mostly bacteria that live on and in every living thing. While bacteria play a major role in human diseases, they can also be useful to people. Cheese and yogurt are made by the addition of bacteria.

Bacteria serve as food for small organisms, which in turn are eaten by larger animals. Finally some of the animals fed by bacteria become part of the food chain that includes humans. For example, many crustaceans or shellfish eat materials worked upon by bacteria. Shellfish is a food consumed by humans.

1. Name the five main kingdoms of the living world.
2. Name two differences between plants and animals.
3. Why do scientists put protists, fungi, and monerans into separate kingdoms?
4. What are microorganisms?
5. How are protists similar to both plants and animals?

Summary of Chapter 2

The two main categories of objects in the world are living things and non-living things. Living things, or organisms, can move, breathe, get food, and perform the basic life activities. Non-living things are unable to do these things. Living things are also more more varied and more complex than non-living things.

There are billions of organisms in the world. They are easier to understand if we separate them into categories. Organisms, or living things, are divided into five major kingdoms: plants, animals, protists, fungi, and monerans.

1. *biology*
2. *organism*
3. *natural things*
4. *living things*
5. *non-living things*
6. *basic life activities*
7. *properties*
8. *organ*
9. *kingdoms*
10. *plants*
11. *animals*
12. *microorganisms*
13. *protists*
14. *microscope*
15. *bacteria*
16. *protozoans*
17. *fungi*
18. *absorb*
19. *parasites*
20. *decompose*

REVIEW QUESTIONS FOR CHAPTER 2

1. Name the two main categories in the world of natural things.
2. Of what three basic materials are non-living things made?
3. Name the properties of this book.
4. Besides basic life activities, what else makes living things different from non-living things?
5. List the five major groups of organisms in the world.
6. What are some characteristics of each of these five groups?
7. Name three ways bacteria can be helpful to humans
8. How can bacteria be harmful to humans?
9. What is a parasite?
10. What is the most important job of the fungi?

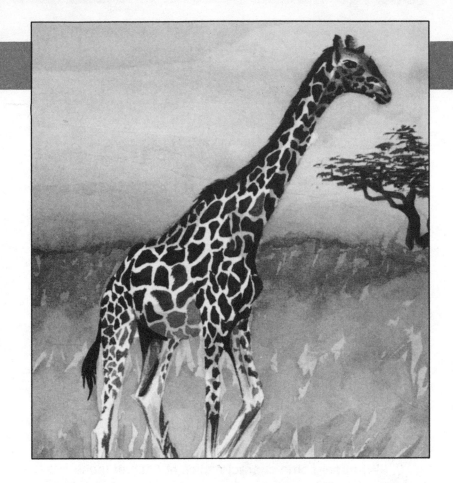

CHAPTER GOALS:

To learn how to use a system for classifying and organizing animals.

To use a system of classification for organizing and identifying some familiar animal groups.

KEY IDEAS:

Animals are classified into different groups, according to their similar structures.

Each animal has its own special, two-word name.

The most visible animals in the animal kingdom are animals with backbones. They are called vertebrates.

Animals without backbones make up a major group in the animal kingdom. They are called invertebrates.

CHAPTER 3

Classifying and Organizing Animal Groups

Introduction

There is a system of **classification** for all animals by names and categories. An example of how the system works can be seen in the case on the shoebill stork, a large wading bird. This very large bird is a relative of other storks and herons. All of the wading birds in the world have similar structures that are useful to their life in and around water. Within the whole group, each one of the wading birds has structures that separate it from other kinds of wading birds. The shoebill is so different from its relatives that it is called "one of a kind" by the scientists who study birds. The shoebill has a special scientific name that identifies it. The shoebill is called *Balaeniceps rex*. It is

Shoebill stork — A wading bird

Balaeniceps rex

the only bird in the world with that name. It means "shoe-bill stork."

In the classification system for organizing animals there is a special name for each kind of animal. This chapter is about that naming system. You will see why the naming system is useful and important for everyone.

Using Similar Structures to Classify

Birds of a Feather

It is easy to classify birds as a single group of animals because only birds have **feathers.** No other animals have feathers. Feathers are used for flying and to conserve or keep in the body heat of birds. One single structure, feathers, separates the birds of the entire world into a group apart. To this group belong robins, sparrows, mockingbirds, bluejays, crows, starlings, ducks, and brown-headed cowbirds.

Raccoon
Procyon lotor

Animals with Fur

Another large group of animals are those with **fur** or hair. Animals like bears, raccoons, squirrels, deer, lions, tigers, rabbits, rats, cats, dogs, elephants, orangutans, and giraffes have fur. Fur helps insulate the body and keeps it from losing body heat. In the system of classification, animals that have fur and feed their babies milk are called mammals. Some of the common mammals are dogs, cats, squirrels, whales, bears, and humans.

Lion
Felis leo

Scales

Scales are the outside body covering of fish. If you have ever scaled a fish, you know about the hundreds and hundreds of small scales that are on the outside of the fish's body. Scales are part of a fish. Bass, trout, sunfish, perch, pickerel, tuna, salmon, goldfish, carp, and guppies are fish, and all have scales.

But snakes, lizards, and turtles have scales, too! So, having scales as a body structure does not mean "for fish only!" Snakes, lizards, and turtles are reptiles. How can we separate the reptiles from the fish? We need to find some separate structures for each kind of organism. To classify into separate groups requires identifying unique traits.

Scientists have found that fish have hearts that have one chamber. Reptiles have hearts with two chambers. Also, fish must live in water. Reptiles may live on land as well as in the water. In addition, reptiles lay eggs with soft shells. Scales alone are not useful in separating fish from reptiles. So two large groups of animals—fish and reptiles—must be separated in other ways.

CHECK YOUR UNDERSTANDING

1. What body structure do birds have in common?
2. What other body structures are used to classify animals into groups?
3. What animals have scales?
4. What characteristics do scientists use to separate fish and reptiles into different groups?

KEY IDEA #2:

Each animal has its own special two-word name.

All Animals Fit Somewhere in the System

Each kind of animal has a place of its own in the system of classification. The system of names keeps track of thousands and thousands of animals. The system

---

works on a simple plan of different *levels*. The highest level is the level of kingdom. Once an organism is placed in the **animal kingdom**, then the scientist needs to decide the kind of an animal it is.

The scientist asks questions about the structure of the animal. The information is narrowed down to the two lowest levels, **genus** and **species**. Each animal receives its scientific name of two words. These names are in the Latin language and no other animal can have that name. The words name the genus and the species of the animal. The system is called binomial nomenclature. This means two names.

Scientific Names

In the case of the human animal, the scientific name is *Homo sapiens*. The first word tells what genus a human being is. The second word tells to what species a human belongs. Whenever you see the scientific name *Homo sapiens*, you can be sure that this is another name for "human." *Homo sapiens* is Latin for "wise man."

The common pet that you call "dog" is in the animal kingdom. Its scientific name is *Canis familiaris*. It is in the genus *Canis*. Its species name is *familiaris*. In the Latin language it means "familiar canine."

The little house mouse is known to scientists as *Mus musculus*. It is in the genus *Mus*. Its species name is *musculus*. There is no other animal in the world by that name. If you find another animal that is very much like a house mouse, but is not really a house mouse, you will have to give it another name.

House mouse
Mus musculus

Basic Biology

The Seven Levels of Classification

The chart below shows the seven levels that scientists use to classify each animal. These levels are the main categories or groups. Notice that the highest or top level is kingdom. The lowest levels, 6 and 7, are genus and species.

Scientific Names

Homo sapiens

Canis familiaris

LEVELS OF CLASSIFICATION USED WITH THREE ANIMALS			
Levels	Humans	Dogs	House Mouse
1. KINGDOM	Animal	Animal	Animal
2. PHYLUM	Chordate	Chordate	Chordate
3. CLASS	Mammals	Mammals	Mammals
4. ORDER	Primate	Carnivore	Rodents
5. FAMILY	Hominoid	Canid	Murine
6. GENUS	*Homo*	*Canis*	*Mus*
7. SPECIES	*sapiens*	*familiaris*	*musculus*

Three animals are classified on the chart. Look down the column for humans. Look at levels 6 and 7. Can you find the scientific name, *Homo sapiens*?

Find the scientific name for dogs. Find the scientific name for the house mouse.

Read all the way across level 1. All three are animals. Read across level 2. You see that all three are chordates (animals with backbones). Level 3 shows that all three are mammals. They all have fur and feed their young with milk. From level 4 to 7 each animal is completely different. Humans, dogs, and the house mouse are indeed separate kinds of animals. They must each have their own name.

That is the way the system works. Remember that the classification system is based upon **similar structures**. It is the amount of similarity in structure that puts animals together at a level of the system.

1. Name the highest or top level in the system of classification.

2. What do scientists call the first part of an animal's scientific name?

3. What do scientists call the second part of an animal's scientific name?

4. Explain the meaning of this sentence: "Unique or special structures is the key idea in classifying animals."

KEY IDEA #3:

The most visible animals in the animal kingdom are animals with backbones. These animals are called **vertebrates**.

Animals With Backbones

Giraffe

Giraffa camelopardalis

Many animals are similar because they have an internal skeleton made of bones. Human beings have a skeleton, and so do dogs, cats, rats, bats, and elephants. From leftovers at mealtimes we often find bones from chicken, turkey, or fish. These bones have come from the internal skeleton of the animal. The house mouse also has an internal skeleton. So do snakes, giraffes and crocodiles.

Vertebrates

An important part of an animal's skeleton is the backbone. The backbone is made up of a set of bones called vertebrae. The group of animals with vertebrae are called vertebrates. The vertebrates include the most advanced forms of life on earth.

There are more than 50,000 species of animals that have backbones. You have already studied some of the species of vertebrates: *Homo sapiens*, *Mus musculus*, *Canis familiaris*, and *Balaeniceps rex*.

Most of the animals at the zoo are vertebrates. The giraffe is a vertebrate. Reptiles, including lizards, snakes, turtles, alligators, and crocodiles, have backbones and are called vertebrates.

Fish belong to another large group of vertebrates. So do amphibians like toads, frogs, and salamanders. All birds have a spinal cord made up of vertebrae. They, too, are called vertebrates.

Classes of Vertebrates

The vertebrates are not all alike, even though they all have internal skeletons made up of bones. In the system of classification vertebrates fall into five separate **classes.** Fish, amphibians, reptiles, birds, and mammals each belong to a separate class. Each class has distinct characteristics.

The chart below shows how the leopard frog and the box turtle are classified. On the left you see the seven levels of the classification system. Look at level 2. Find the word chordate two times. Chordate means that the animal has a backbone. Both the leopard frog and the box turtle have backbones. Therefore, both animals are vertebrates.

LEVELS OF CLASSIFICATION OF TWO ANIMALS		
Levels	Leopard Frog	Box Turtle
1. KINGDOM	Animal	Animal
2. PHYLUM	Chordate	Chordate
3. CLASS	Amphibian	Reptile
4. ORDER	Salientia*	Chelonia**
5. FAMILY	Ranid	Emydid
6. GENUS	*Rana*	*terrapene*
7. SPECIES	*pipiens*	*carolina*

* *Salientia* means "to leap, bound, or jump"
** *Chelonia* means "tortoise or turtle"

Rana pipiens

Terrapene carolina

The next level, level 3 tells you that the leopard frog is an amphibian, and that the box turtle is a reptile. Levels 6 and 7 give the special scientific names—the genus and the species. The scientific name of the leopard frog is *Rana Pipiens*. The box turtle's scientific name is *Terrapene carolina*.

Animals are classified or grouped together by their similarities. The system lets you see that each kind of animal is different enough to have its own name.

CHECK YOUR UNDERSTANDING

1. What is the major similarity of structure in the animals known as vertebrates?
2. About how many species of animals with backbones are there?
3. What are the five classes of vertebrates?
4. What is the genus of the leopard frog?
5. What is the species of the leopard frog?

KEY IDEA #4:

Animals without backbones make up a major group in the animal kingdom. They are called invertebrates.

Animals Without Backbones

The animal kingdom is divided into two large groups—animals with backbones, the vertebrates; and animals without backbones, the **invertebrates**.

The invertebrates outnumber the vertebrates by far. There are more than one million different species without backbones. There are so many differences among the invertebrates that scientists have classified them into eight separate groups. Each of these groups has a large number of species. For example, in the group known as the **sponges**, there are 5,000 species.

Roll Call of the Invertebrates

Here is a roll call and a very brief description of the groups of animals known as invertebrates. These animals do not have backbones.

1. Sponges (Porifera)

Sponges are the simplest invertebrates. They live mostly in salt water. They have only two layers of cells in their body wall. Sponges do not even look like animals. They are attached to one spot much like plants are anchored to one spot. You may be familiar with the bath sponge, which is really the "skeleton" of a sponge.

Sponge

2. Jellyfish, Sea "Flowers," and Coral (Cnidaria)

There are nearly 10,000 species of these invertebrates in the world. Most species live in the sea. These animals have tentacles with stinging cells. Some live alone, but others, like the Portuguese man-of-war, live in colonies. These organisms float about for the most part. As they float along, they "trap" their food. The corals, however, stay in one place. A fresh-water relative of the jellyfish is the hydra.

Jellyfish

3. Flatworms (Platyhelminthes)

Flatworms have flat bodies. Most flatworms are parasites. They live inside an animal host. The most famous parasites are the tapeworm and the liver fluke. Tapeworms live as parasites in the large intestines of humans. They live the life of "highway robbers," taking food that does not belong to them. To be free of tapeworms, humans should eat only well-cooked meat.

Planaria
15,000 species

A fresh-water flatworm, the planaria, is interesting. If it is cut in two or more pieces, each piece will grow into a new planaria. This special kind of regrowth is called regeneration.

4. Roundworms (Nematoda)

There are more than 12,000 species of these tiny animals in the world. The hookworm is found in soil and

Hookworm

water, especially where sanitation is poor. Humans may pick up hookworms through their bare feet in unsanitary places. The roundworm known as the vinegar eel can be found in undistilled cider vinegar.

5. Segmented Worms (Annelida)

Earthworm

There are about 10,000 species of **segmented** worms, or worms with segments in their bodies. These worms include the common earthworm (night crawler) and the leeches. Earthworms perform the useful job of tunneling their way through soil and bringing materials to the surface. The burrows or tunnels that they make let air flow into the soil. Leeches are best known as blood-sucking parasites. They were once used by doctors to draw blood from patients in order to reduce a fever.

6. Starfish and Other Spiny-Skinned Animals (Echinodermata)

The 6,000 species in this marine group include sea urchins, sand dollars, and sea cucumbers. **Starfish** are the most well-known. Starfish have tube feet on the underside of their arms that act as suction cups to hold tight to clam shells. Starfish force the clam shells open, invert their stomach into the shell, and eat the clam. Some common names are misleading. The *starfish*, for example, is no fish at all.

Starfish

7. Mollusks (Mollusca)

Snail

There are 100,000 species of **mollusks**, including the octopus, squid, clam, and snail. They live in the sea, on land, and in fresh water. Land snails, pond snails, and garden snails are the largest number of species in the mollusk group. Most species have shells and soft bodies. Shells are either in one piece, like the land snails, or in two halves, like the clam shell. Clams are bivalves because of their two shells. Most mollusks have a muscular foot that they use to move around.

Squids and octopuses are special mollusks because they have a powerful nervous system. They move very rapidly and do not have a shell like other mollusks. Squids and octopuses are called cephalopods. Cephalopods eat other animals. They are carnivorous. These unusual mollusks are able to change their color to match their environment.

8. Arthropods (Arthropoda)

Arthropods are by far the largest group of invertebrates. There are more species of arthropods than any other single group of animals in the world. Nearly 1,000,000 species are known, and more are counted each year. The arthropods are given much attention because of their great variety, their large number of species, and because they have a special high level of interest to humans. The word *arthropod* comes from the Greek language. It has a special descriptive meaning. *Arthro-* means "joint." The word *-pod* means "foot." Put them together and you have "joint foot." There are five classes of arthropods: crustaceans, insects, arachnids, centipedes, and millipedes.

Similar Structures of the Arthropods

Similar structures among the arthropods are 1) jointed legs, 2) claws with joints, and 3) sectioned or jointed bodies. The arthropods also have exoskeletons, or outer body coverings. The outer covering varies from very hard, as in crabs and other crustaceans, to hard, as in certain insects and millipedes.

Hardshell crab

The hardshell crab is a good example of all of these similar structures. Its legs, claws, and body parts are all in sections that are jointed. Crabs shed their exoskeletons as they grow. Each time they shed, they grow a new exoskeleton until finally they become adult crabs. Arthropods in general follow the same pattern. A familiar sight is the shell, or exoskeleton, of a cicada, or locust, hanging on the bark of a tree. These arthropods go on to grow new exoskeletons which they will also shed.

Ladybird beetle

Mosquito

The Insects

More than 900,000 kinds of insects have been identified. Some are wingless, but most have one or two pairs of wings. All insects have a head, a thorax, and an abdomen.

Insects have a pair of antennae on their head. The thorax, or middle part, of their three-part body has three pairs of legs attached. Most insects have two pairs of wings that are also attached to the thorax. The insects are a very successful group of animals. They occupy just about every kind of living space that there is: in and on the water, in trees and on plants, on the ground and under the surface, in old logs, and in most of the places where humans live. Many are pests. They can destroy crops. Some insects, however, serve as friends of humans by eating other insect pests. Insects live everywhere except in the oceans.

Horseshoe crab

Arachnids and Crustaceans

The most important relatives of the insects are the arachnids and the crustaceans. The arachnids include spiders, daddy-long-legs, and horseshoe crabs. The horseshoe crab is not a crab at all. It is more closely related to spiders. The crustaceans are the crabs, lobsters, barnacles, sowbugs, and crayfish.

Lobster

Centipedes and Millipedes

The story of the arthropods would not be complete without including the *centipedes* and the *millipedes*. The

Centipede   Millipede

centipedes have one pair of jointed legs on each of their body section, or segments. They are predators of insects. Predators are animals that capture other animals as their main source of food. Centipedes "prey upon" insects. The millipedes have two jointed pairs of legs on each of their body sections or segments. A common name for the millipede is "thousand-legger." Millipedes are important forest *decomposers*. By breaking down the larger materials that fall on the forest floor, they make the soil more fertile.

ARTHROPODS — ANIMALS WITH JOINTED LEGS
The highest number of species on the face of the earth.

Classification and Number	Legs	Examples of Members
Insects 900,000 species	3 pairs of legs	crickets, ants, bees, houseflies, grasshoppers, butterflies
Arachnids 55,000 species	4 pairs of legs	spiders, mites, horseshoe "crab," ticks
Crustaceans 30,000 species	5 pairs of legs	lobsters, crabs, shrimp, crayfish
Centipedes 2,500 species	1 pair of legs per section	house centipede, garden centipede
Millipedes "Thousand-leggers" 10,000 species	2 pairs of legs per section	garden millipede, meadow millipede

CHECK YOUR UNDERSTANDING

1. What are the two main "similar structures" of the group known as arthropods?
2. What is the meaning of the term *arthropod*?
3. Name the five classes of arthropods.
4. Which of the arthropod groups has the largest number of known species? About how many?
5. What is the major difference between millipedes and centipedes?

Summary of Chapter 3

Scientists use a classification system for organizing animals. There are seven levels of classification. Animals are grouped and classified according to their similar structures. When an animal is classified, it is given a scientific name of two parts or two words—a genus and a species.

Two major groups of animals in the world are the vertebrates, animals with backbones, and the invertebrates, animals without backbones. There are eight different groups, or major kinds, of invertebrates. The largest group of animals within the invertebrates is the arthropods. Arthropods have skeletons and jointed legs.

1. *wading bird*
2. *classification*
3. *feathers*
4. *fur*
5. *mammals*
6. *scales*
7. *reptiles*
8. *animal kingdom*
9. *genus*
10. *species*
11. *chordate*
12. *similar structure*
13. *vertebrate*
14. *classes*
15. *invertebrate*
16. *sponges*
17. *segmented*
18. *starfish*
19. *mollusk*
20. *arthropods*

REVIEW QUESTIONS FOR CHAPTER 3

1. What is the "similar structure" that is used to classify birds of the world into a single group?

2. Raccoons, tigers, cats, dogs, and giraffes are classified into a large group. What is the "similar structure" that is shared by these animals?

3. *Canis familiaris* is the scientific name for a dog. Which part of the name is the genus? Which part is the species?

4. Name the seven levels used in the system of classification.

5. What are some animals that have an internal skeleton?

6. How do animals qualify as "vertebrates?"

7. Name the five different classes of vertebrates.

8. What is the chief "similar structure" of the invertebrates?

9. What are the eight major groups of invertebrates?

10. Name the five classes of arthropods.

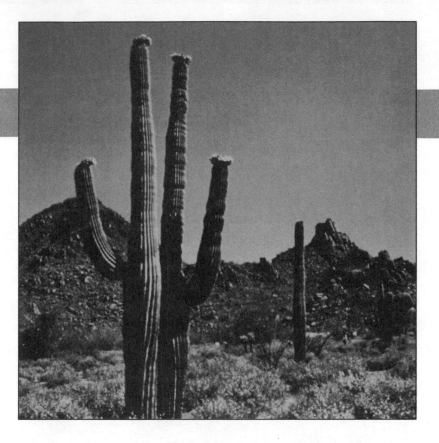

To understand that plants are grouped and classified according to their similar structures.

To know how to use the two-word scientific classification system for names of plants.

KEY IDEAS:

Plants are classified into different groups, according to their similar structures.

Seed plants include flowering and non-flowering plants with vascular systems or tube systems that extend from roots to leaves.

The ferns are vascular plants with roots, stems, and leaves. Ferns use spores as a means of reproduction.

Mosses are non-vascular plants that reproduce by means of spores.

4

Classifying and Organizing Plant Groups

Introduction

The most familiar parts of the world of living things are divided into two divisions: the animal kingdom and the plant kingdom. You might be tempted to ask, "Which of these kingdoms is more important?"

The answer would have to be "The plant kingdom." Plants could survive without animals, but animals could not survive without plants. Plants have the ability to make their own food. Animals, however, need plants for their food.

KEY IDEA #1:
Plants are classified into different groups, according to their similar structures.

Classifying Plants

The same kind of two-name system that you used for the classification of animals is useful in classifying plants. Plants are classified according the the presence or absence of certain body parts, such as seeds, conducting tubes, and roots, stems, and leaves. The three main groups of plants are the seed plants, the ferns, and the mosses.

Botanists

The system of plant classification was invented in 1753. **Carolus Linnaeus**, a Swede, gave names to all of the known plants and animals in Europe. There are some scientists who have spent their entire adult lives studying and classifying plants. These scientists, called botanists, have traveled around the world in search of new species of plants. By describing the structures of plants and the places where they are found, botanists contribute to the science of **botany**. Botany, the study of plants, is one of the major branches of the science of biology.

Plants with Transport Systems

The seed plants and ferns are the major groups of plants that have an internal system for transporting food and water. They are called **vascular plants**. Vascular plants have conducting tubes called vascular tissue. Vascular tissue is made of tube-like structures. The tubes are connected into a network from the roots to the stems and into the leaves. Vascular tissue helps plants by providing a pathway for water and food to be moved to and from different parts of the plant body. For example, the roots of plants pick up water and minerals from the soil and transport them to other parts of the plant.

The vascular system is important is two ways. It allows the plant to grow tall. It also allows plants to grow in places where water is not always present.

Plants without Transport Tubes

In contrast to the vascular plants, there is a large group known as the **non-vascular plants**. These non-vascular plants must be in constant contact with moisture. They don't have tubes for transport of food and water and they do not grow very tall. These small plants grow in damp places on the ground and on the sides of trees and rocks. They are called mosses.

CHECK YOUR UNDERSTANDING

1. What are the three major groups of plants?

2. Why are green plants so important to animals?

3. Name some "similar structures" of plants.

4. What are the differences between vascular and non-vascular plants?

5. Name of the special branch of biology that deals with the study of plants.

KEY IDEA #2:

Seed plants include flowering and non-flowering plants with vascular systems.

The Seed Plants

The seed plants are the largest group of plant species in the world. There are over 270,000 species. Seeds contain specialized reproductive bodies which make these plants more successful than ferns or mosses.

Seed plants range in size from the smallest flowering plant, the duckweed, to the largest plants in the world, the giant redwoods, sequoias, and oak trees.

Seed plants have great variety and live almost everywhere. Grass, trees, garden flowers, roadside weeds, pine trees, bushes, vines of all kinds, flower-shop flowers, and cactuses are all seed plants. Also, oats, corn, wheat, rice, cotton, and barley are all plants that grow from seed. The seed plants have vascular tissues. They are divided into two sub-groups—flowering plants and non-flowering plants.

Flowering plants

There are about 250,000 species of flowering plants. The word used to describe this large group is **angiosperm.** This word was made from the Greek words *angeion*, "capsule," and *sperma*, "seed." Seeds have "seed coats" that act like small containers or capsules.

Inside of the seed capsule is the stuff to begin a new plant. Angiosperms have their seeds covered. Angiosperms come in two kinds of seed forms called monocots and dicots. Flowering plants are either one or the other.

Dicots

There are more than 165,000 species of dicots. They have two "seed leaves" or embryo leaves inside the two halves of the seed coat. Bean seeds have two halves, and so do peanuts. You are probably most familiar with the two halves of an individual peanut.

Embryo leaves

Dicot (bean)
Two seed coats

Branching veins

The lima bean is a dicot plant that shows the structure clearly. The seed has two halves. When the seed is opened, you can see the embryo leaves. Also, when you grow bean plants from the seeds, you will see that the leaves have a network of veins rather than a set of **parallel** veins.

Some other dicots are trees that produce wood for furniture and trees that are used as shade trees. Humans use dicots in the form of fruits and vegetables.

Parallel veins

Monocot (corn)
One seed coat

Embryo leaf

Monocots

There are more than 60,000 species of monocots. They include corn, wheat, and rice. Grasses that cattle feed upon are also monocots. Monocot seeds have only one, single embryo leaf. Look in a grain of corn for an embryo leaf. Monocots have leaves with veins that run parallel. You can see parallel veins in a blade of grass, for example.

Non-Flowering Plants

The non-flowering plants are called **gymnosperms**. There are about 700 species. The word *gymnosperm* means "naked seed." Seeds in the gymnosperms are described that way because they are not enclosed in a fruit such as the apple. Instead, gymnosperm seeds are produced inside of cones. The major group of gymnosperms are the **conifers**. *Conifer* means "cone-bearing." Gymnosperms do not produce flowers.

Conifer

The conifers make up 600 of the 700 species of gymnosperms. Included are the pines, spruces, and fir trees. The giant redwoods and the tall growing spruces are among the conifers that make up a major part of the forests. Christmas trees are most often spruces, firs, and pines. Decorative plants such as juniper, yews, and spruces are familiar plants in the landscape of many homes.

Conifers have needles. The needles are the leaves of the conifers. They are able to hold water more easily than broad-leafed trees. This makes it easier for them to live in dry and sandy places where the tree is forced to store water for longer periods. Conifers have leaves all year. For this reason they are called evergreens. They lose only a part of their leaves at any time. They appear to be green all the time.

Needlelike leaves   Branch

Cone

Pine tree

Conifers grow in places that are too harsh for other plants: along the seashores, in dry areas, and in shallow, rocky soil. You can see the bald cypress in swamps. The wood of the conifers is called **softwood**. It is used as a source of lumber and as pulpwood for making paper.

The Ginkgo Tree

Besides pine, spruce, and fir trees, there are other gymnosperms. The ginkgo tree of city streets is one of the most familiar. They are used to line many city streets because they are able to survive pollution better that other trees. Gingko trees have peculiar fan-shaped leaves and are resistant to disease as well as insects. The gymnosperms are classified according to their similar structures. They have seeds, and most have cones and needle-like leaves. They all have vascular tissue.

Leaves of ginkgo tree

CHECK YOUR UNDERSTANDING

1. What are the scientific names for the flowering plants and the non-flowering cone-bearing plants?
2. What is a seed coat?
3. What are the differences between dicots and monocots? Name two plants that are dicots and two that are monocots.
4. Why are the seeds of conifers called "naked seeds?"
5. List some characteristics of conifer leaves.

KEY IDEA #3:

The ferns are vascular plants with roots, stems, and leaves. Ferns use spores as a means of reproduction.

The Ferns

There are about 10,000 species of ferns in the world. Many of them are tropical plants. **Ferns** are the largest group of vascular plants that do not have seeds. Ferns have **spores**. Spores are the reproductive bodies of ferns. They are found as small spots called sori on the underside of fern leaves. Spores are formed into small clusters that dry up and burst open when they are ripe. There are millions and millions of them, and they are scattered by the wind.

Spores must find a moist place to land and get started. If they don't (and millions upon millions don't), they dry up. Seeds, on the other hand, carry moisture and food inside of their outer seed coats. The protective seed coat allows them to stay alive for a period of time. The seed plants are more successful for this reason.

Spore case

Frond, or leaf, of a fern

Fossil Ferns

At an earlier period in Earth's history the ferns were the main form of plant life. Some 200,000 million years ago they were the size of trees. Some of them became the coal deposits that are mined in different parts of the world today.

Frond

Young leaf

Rhizoid

Roots

A fern

Ferns as Decoration

Ferns are used as decorative plants because of their leaves. The leaves are large and flat, and the patterns of the leaflets are attractive. Probably most of the ferns that you have seen have been potted ones. Rhizoids or runners are underground along with the roots. A stem-like midrib supports the leaflets. They are like the seed plants in having roots, stems, and leaves.

CHECK YOUR UNDERSTANDING

1. Compare and contrast ferns and seed plants.
2. How are ferns and seed plants different?
3. How do spores function to reproduce new plants?
4. Why are the seed plants more successful than the ferns?
5. What became of the large fern plants of long ago?

KEY IDEA #4:

Mosses are non-vascular plants that reproduce by means of spores.

The Mosses

Scientists have found about 14,000 species of mosses. A **moss** is a non-vascular plant. They have no tubes to transport water. Therefore, they must live in moist, shady places.

A close look at a moss reveals leaf-like and stem-like parts. They look like little trees. They often form carpet-like mats where the soil and the surrounding air are moist. Woodlands and the edges of streams are common homes for mosses. Air full of moisture and soil that is wet provide the best home, or habitat, for the moss plants.

Moss Reproduction

Mosses reproduce by means of spores. Millions of tiny spores form inside spore cases on special stalks or "stems." The spore case breaks open when it is ripe. It shoots the moss spores into the air. The spores make new plants when they fall on some moist soil.

Reproduction in the mosses depends on the presence of water. The reproductive sperms need to "swim" to the egg. The resulting embryo grows into a stalk that has the spore case at one end.

Not all of the spores find good places to grow. Those that do make new moss plants. One reason for moss survival is that great numbers of spores are produced. Some spores have been known to survive for many years and to go on to produce good moss plants.

Shaggy moss

Mosses Build Soil and Bogs

Mosses have been called "pigmy plants" because of their small size and their resemblance to tree plants. They are able to start growing in places where other plants cannot take root. In thin soils, cracks in rocks, and within the bark of fallen trees are some of the places for mosses to grow. Mosses help in the formation of soil. When moss plants die, they form humus. Humus helps make good soil.

Mosses of the genus *sphagnum* are known as bog builders. A **bog** is wet, spongy ground that is formed from decayed moss and other plant matter. Peat bogs are made from decayed and compressed sphagnum moss. Peat moss is mixed with soil to improve gardens. Peat is dehydrated and used a fuel in Europe.

1. What does "non-vascular" mean?
2. Where do mosses, which are non-vascular, need to live?
3. Describe the process used by mosses to reproduce.
4. How do humans use mosses?

Summary of Chapter 4

Plants are organized into three major groups, based on their similar structures, or body parts. The three groups are the seed plants, the ferns, and the mosses.

Seed plants and ferns are vascular plants, with tubes that transport water and food from the roots to the other parts of the plants. Of the two groups of vascular plants, seed plants are the more numerous. They also have more species and greater variety. Seed plants reproduce by means of their seeds.

Most of the seed plants are angiosperms, or flowering plants. Angiosperms are either monocots with one embryo leaf, or dicots with two seed leaves. The majority of plants are angiosperms.

The other seed plants are gymnosperms, or non-flowering plants. Gymnosperm seeds are not enclosed in a fruit, but are found in a cone. Many gymnosperms are conifers.

Another group of vascular plants is the fern group. Although ferns have vascular tissues, they do not have seeds. Instead, they reproduce by means of spores.

Mosses are non-vascular plants. Like ferns, they reproduce by spores. Mosses help in building soil and bogs. They also provide humus which enriches the soil.

1. *transport tube*
2. *botanist*
3. *botany*
4. *vascular*
5. *non-vascular*
6. *Carolus Linnaeus*
7. *angiosperm*
8. *gymnosperm*
9. *reproduction*
10. *dicot*
11. *monocot*
12. *parallel*
13. *seed*
14. *embryo*
15. *conifer*
16. *softwood*
17. *fern*
18. *spore*
19. *moss*
20. *bog*

REVIEW QUESTIONS FOR CHAPTER 3

1. What is the major relationship between plants and animals?

2. What groups of plants reproduce with spores instead of seeds?

3. What are some of the "body parts" or characteristics used to classify plants?

4. How does vascular tissue work in the life of plants?

5. What is botany?

6. What group of plants has the highest number of species?

7. What are the scientific names for flowering and non-flowering plants?

8. Why are conifers able to live in harsh places? Name four trees that are conifers.

9. Why are conifers called evergreens?

10. List three ways that fern reproduction differs from reproduction of seed plants.

11. List three reasons mosses need moisture to stay alive.

12. What is the best habitat for mosses?

13. Why is the ginkgo tree used to decorate city streets?

14. How do mosses reproduce?

15. What are bogs? Tell how a bog is formed.

To describe the basic life activities of the protists, fungi, and monerans.

To recognize the great variety of simple forms of life in the kingdom of protists, the kingdom of fungi, and the kingdom of the monerans.

KEY IDEAS:

Protists are protozoa and algae. They are simple organisms that belong to a third kingdom, separate from plants and animals.

Protists carry out basic life activities and affect the lives of other organisms.

Fungi are a group of organisms that lack chlorophyll, reproduce by spores, and belong to a fourth kingdom, separate from plants, animals, and protists.

CHAPTER 5

Organizing Protists, Fungi, and Monerans

Introduction

For a long time, keeping track of and identifying organisms was based on the answer to the question, "Is it a plant or is it an animal?"

Organisms have now been discovered that do not fit neatly into the category of plant or animal. As far back as 1866, one scientist suggested that a third kingdom called *Protista* should be formed to include organisms that were one-celled. So the kingdom *Protista* was organized.

Protists have special traits. They are mostly microscopic organisms such as protozoa and algae. Protozoa and algae can show both plant-like and animal-like characteristics. They deserve a complete kingdom level of classification all their own.

Another kingdom called **fungi** was added later because they have separate traits. Fungi are organisms such as yeasts or molds that are rooted to one spot the way plants are, but they cannot make their own food because they lack chlorophyll. So, many fungi live off dead or decaying material.

Most recently, biologists have used characteristics of monerans to separate them into a single kingdom.

Monerans are now the fifth major kingdom—the kingdom of organisms without nuclei.

In keeping with current knowledge, the old categories of plant and animal have been expanded to five kingdoms. In this textbook, you will study the five kingdoms of *plants, animals, protists, fungi,* and *monera.*

Protists, fungi, and monera are classified and organized under the same two-name system used for plants and animals. In each of those kingdoms, there is only one organism that can be called by the name of its genus and species. For example, in the kingdom of the protists there is only one *Euglena gracilis.* The genus is *Euglena* and the species is *gracilis.*

A protist, *Euglena gracilis*

KEY IDEA #1:

Protists are protozoa and algae. They are simple organisms that belong to a third kingdom, separate from plants and animals.

The Protist Kingdom

The protists include some animal-like and some plant-like organisms. Animal-like protists, such as the ameba, paramecium, and euglena, move around and capture food. **Algae** are plant-like and are able to make their own food. The euglena is also plant-like because it can make some of its own food. The euglena has a flagellum, or whip-like tail, that helps it move.

Basic Biology

Plant-Like Protists

Life on earth would not be the same without the plant-like *protists*. They make their own food and are food for thousands of animal species in the seas. Most are microscopic, and live in both salt and fresh water. The plant-like protists float and swim near the surface of the oceans. They reproduce rapidly. Their numbers are measured in the billions and trillions. One gallon of sea water taken from the surface of the ocean would hold many thousands of protists. The protists provide the basic food source for all of the animals of the seas.

When plant-like protists make food, they carry out the process known as photosynthesis. During photosynthesis, the organisms give off oxygen to the atmosphere. The oceans of the world cover three-fourths of the earth's surface and are the home of the greatest oxygen producers — the protists. It has been estimated that these microscopic organisms provide 50% of the earth's atmospheric oxygen.

Two big jobs done by billions of these little-known workers are feeding ocean animals and providing oxygen for other life forms.

Algae—Plant-Like Protists

The algae are mostly aquatic protists. They live in water. There are more than 20,000 species of algae.

1. Spirogyra

Spirogyra is the green algae found in ponds. It is more commonly known as pond scum. This bright green protist reproduces rapidly and forms floating masses on pond surfaces. There are more than 60 different species of Spirogyra. All of the different Spirogyra have a strand, or string, of **chloroplasts**. This green material contains **chlorophyll**, which is useful in the energy-producing process of photosynthesis.

2. Protococcus

Another genus of green algae, *Protococcus*, is very common. ***Protococcus*** is a land organism, unlike most of the other green algae. *Protococcus* is found growing on the bark of the shaded sides of trees. It also grows in moist places like damp stones and damp fence posts. The next time you walk in the woods, see if you can find this green algae on the shaded and moist side of the tree trunks.

3. Sea Lettuce

Sea lettuce is a marine alga that is seen at the seashore. It is called that because it looks like regular lettuce leaves. Common garden salad lettuce is in the genus *Lactuca*, and because of the strong resemblance in appearance, sea lettuce has the scientific name, *Ulva lactuca*. Notice *lactuca* in the names for both organisms. Sea lettuce may grow to six inches wide and two feet long. Sometimes after a storm, sea lettuce fouls the shore and causes a bad odor. Sea lettuce is found along the coasts of the Atlantic and Pacific Oceans.

4. Brown Algae

The giant kelp of the Pacific coast is reported to be the longest plant in the world. It grows along rocky shorelines. This huge **brown algae** called *Fucus* grows up to 210 feet long and is supported by bladders to help it float. The plant body contains iodine and a substance called *algin*, which is used to make foods firm and uniform. This plant is harvested in California, much the same as a corn crop is harvested in the midwest.

5. Diatoms

Diatoms are protists that are the most numerous in the 6,600 species of brown or gold algae. They are easy to find in both fresh and sea water. They have a great variety of shapes. The shells of these microscopic organisms are made of silicon, the same material used to make glass.

Diatoms grow in large numbers, and are a major source of food for many fish. The shells make silica deposits on the ocean floor. Since diatoms reproduce in a matter of days, each new reproduction period results in piles of these deposits. The "ooze" at the bottom of the sea is made of these remains. This ooze is mined to make metal polish, toothpaste, soaps, and scouring powders.

Animal-Like Protists

Animal-like protists are called protozoa and can move on their own. They cannot make their own food. For these reasons, they resemble animals more than plants. Plants are generally stationary and make their own food.

1. Ameba

The most familiar of the protists is the ameba. The way the ameba moves is its main feature. The ameba makes "false feet" by flowing along. It also takes in food with its "false feet." Cytoplasm, or watery insides, pushes out the covering or cell membrane. Amebas live on surfaces of rocks and plants in ponds. The ameba is one-celled. It belongs to a group known as the protozoans. *Proto-* means "first." *Zoa* means "animal." A two-word scientific name for one of these organisms is *Ameba proteus.*

An ameba

2. Paramecium

Probably the second most popular protist organism in the world of biology is the paramecium. Parameciums move around to get their food and to respond to trouble in their environment. Because they move by means of **cilia**, or hair-like structures, they are called *ciliates*. Sometimes the cilia cover the entire outside of the body.

Paramecium caudatum is a common pond inhabitant. It keeps a more organized shape than the ameba. It is usually seen moving along front end first, but it does have a reverse gear, and can back up. It is shaped like a slipper.

3. Trypanosoma

Trypanosoma are animal-like protists with a tail-like flagellum that moves them through the blood of animals. One species lives part of its life in the tsetse fly in Africa. It causes African sleeping sickness in humans.

4. Sporozoans

Sporozoans are animal-like protists. They are all parasites and as adults have no means of moving about. They reproduce by means of spores. One sporozoan, *Plasmodium vivax*, is famous because it is the cause of malaria in humans. Malaria is a disease of tropical countries and continues to be a problem disease for millions of people. This sporozoan is transferred to humans by the bite of a mosquito called *Anopheles*.

5. Euglena

Euglena gracilis is almost evenly divided between being plant-like and animal-like. It is plant-like because it has chloroplasts and chlorophyll. It is able to make its own food. Euglena is much like an animal because it uses a whip-like flagellum to move through the water. **Euglena** is a one-celled organism and fits the description of the protozoa. It has only a cell membrane like other animals; it does not have the typical cell wall that plants have. Euglena has a regular shape, but it is very flexible. It can form different shapes at times.

CHECK YOUR UNDERSTANDING

1. Name the five kingdoms used in this book.
2. What are the names of three plant-like protists?
3. What are the names of three animal-like protists?
4. What do the plant-like protists do for the animals of the sea and the world?
5. What part do chloroplasts and their chlorophyll play in the life of the plant-like protists?

KEY IDEA #2:
Protists carry out basic life activities and affect the lives of other organisms.

Basic Life Activities of Protists

Getting Food

The ameba, paramecium, and the euglena are microscopic organisms that feed on other microscopic organisms. Each of these protists has a special way of getting food.

The ameba moves about by the flow of its "false feet." As it glides along, its body wraps around algae and other protists that serve as food. The ameba engulfs the "food." The food is trapped inside the ameba, and enzymes digest it. This method of wrapping

The ameba getting food by wrapping around a paramecium.

around food material is a special kind of ingestion. The broken-down food passes into the ameba.

Parameciums seem to slide through the water with the aid of their hair-like cilia. The cilia cover the body of the paramecium. They act like little oars, pushing, pulling, and turning the slipper-shaped paramecium through the water. The paramecium has a front end, a back end, and an opening in one side of its body. This opening is called a **gullet** or oral groove. As the paramecium goes along, food materials are pushed into the gullet by the cilia.

Ingested food passes to the end of the gullet and into a bubble-like space called a **vacuole**. Food vacuoles are the place for digestion of food. The vacuoles move around inside the paramecium's one-cell body, bringing energy to the right places. The euglena captures other organisms and gets energy by making its own food with its chloroplasts.

Digestion and Excretion in the Protists

Amebas, parameciums, and euglenas digest their food in special places called vacuoles. Enzymes break down the food materials. The vacuoles then excrete the waste matter.

Vacuole

Food being digested | Nutrients absorbed by cell | Waste vacuole | Wastes excreted

These protists live in a water environment. To keep a balance between the water inside them and the outside water, they have a simple excretion system. Bubbles called vacuoles collect water inside, and then contract to force the water out of the protist's body. The covering, or "skin," of these simple organisms is built to let certain materials come inside. The covering also lets certain materials escape to the outside. There must be some kind of balance between the inside and outside so that the organism can stay alive. The cells in most organisms carry out this kind of balancing act.

Diffusion

You may wonder about water moving into the body of the protist. Because there are materials other than water on the inside of a paramecium's body, there is less water inside. The greater amounts of water on the outside of the paramecium tend to move to where there is less water on the inside. This process is a common event in living things. It is called diffusion.

Diffusion is the movement of particles from one place to another. Particles of water or other substances that are more plentiful in one place will move to nearby places where they are less abundant. As materials move into and out of cells, they follow this pattern of diffusion.

Circulation in the Protists

The ameba, paramecium, and the euglena are protists with just one cell. Circulation of the cell materials is done by water moving into and out of the cell membrane. The water taken in is full of food and dissolved oxygen. Protists depend on oxygen to stay alive. The oxygen reacts with the food to release energy.

Coordination in the Protists

The protists react to conditions in their water environment. Temperature changes cause them to slow down or speed up their movement. A pinpoint of light will cause some of them to gather together in response. Euglena have an eye spot that is sensitive to light. They move to light and carry on photosynthesis.

Temperature and light are called **stimuli**. The action

of the protist is called a response. Stimulus and response are general features of cells alone and of cells acting together in tissues and organs. The protists are simple organisms, but the materials they are made of is just like the cell materials of higher animals. All parts of their cell body are coordinated by reaction to stimuli.

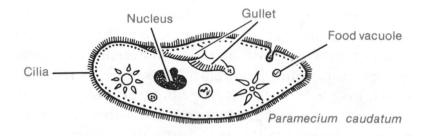

Nucleus Gullet Food vacuole Cilia

Paramecium caudatum

Reproduction in the Protists

Most of the protists reproduce **asexually**. These one-celled organisms simply divide into two **daughter cells**. The daughter cells come about by a process of fission. Each daughter cell receives one half of the nucleus material. For that to be able to happen, the total materials that make up the two daughter cells have to be doubled. The process by which the nucleus divides is called **mitosis**. A daughter cell has the same amount of material in the nucleus as the original protist.

A paramecium dividing

Diatoms live in the warm surface waters of the oceans, where conditions are favorable for reproduction. Diatoms reproduce in a matter of hours. They reproduce asexually most of the time. They are also able to reproduce sexually, at which time the gametes, or sex cells, have a flagellum. Sexual reproduction in the protists, as well as in other organisms, requires gametes that are able to move about.

Sporozoan protists are parasites that reproduce both asexually and sexually. They have complicated life cycles in at least two host organisms. They reproduce by spores at one time and by making gametes at another.

Monerans

The basic organisms in the monerans are non-cellular bacteria. They are among the smallest and oldest forms of life on the earth. There are about 16,000 species of monerans. They grow just about everywhere and act on and with numerous other forms of life.

The Shape and Size of Bacteria

Bacteria are classified mostly by their shapes. Bacteria may be spheres, spirals or rod-shaped. Bacteria are measured in small metric units called microns. It takes a thousand microns to make one millimeter. Typical monerans might measure 2 millionths of a meter.

Basic Life Activities

Even though monerans are so small, they are complete organisms. They carry out basic life activities. Bacteria can feed on dead materials. They break down dead and dying materials on which they land. They use some of the material for their own energy. Most is returned to the soil and reused in the life cycles of plants and animals of future generations.

Respiration in Bacteria

Most bacteria use oxygen and carry out respiration. Bacteria are unusual, however, because some of them do not use oxygen. Those that use oxygen are acting in the same way as more complicated organisms. They use oxygen to release energy. Some use sulfur, carbon dioxide, or nitrogen for respiration.

Reproduction in Bacteria

Reproduction in bacteria is by binary fission. Some bacteria divide every twenty minutes. One common bacteria is *Escherichia coli*, and it divides quickly and often. You can recognize the two-name system at work for even the smallest of organisms. Scientists shorten *Escherichia* to *E.* and call this organism *E. coli*. It is found in the digestive tract of human beings.

Bacteria Are Spoilers

Some bacteria cause food to spoil. When bacteria from the air invade milk, the milk sours. Bacteria grow fastest in warm places. Refrigerating food slows the process of spoilage.

Helpful Bacteria

Bacteria are useful in several ways. They help in the making of several major foods, such as butter, yogurt, and cheese. Bacteria are the organisms used in the study of modern genetics. They are small, reproduce rapidly, and populations of billions are useful in the study of changes that take place in genetic material. Bacteria are also useful in genetic engineering. This fairly new science allows scientists to grow organisms containing specific genes. Bacteria have been engineered to digest waste. Oil spills may be "eaten" by specially designed bacteria that have the ability to digest and change oil.

Nitrogen Fixation and Bacterial Decomposition

Nitrogen is the most plentiful element in the air, but plants are not able to use it directly. Bacteria work in the soil and around the roots of plants. Certain bacteria act on the nitrogen in the air and make it into chemical compounds that plants are able to use. This action of bacteria in the soil is a very important function. Bacteria also act on chemical remains of living materials to break them down. In that way the same nitrogen gets back into the atmosphere. Once back in the air, it is ready to be recycled through the soil bacteria into plants. Bacteria also produce a large supply of carbon dioxide needed by plants for photosynthesis.

CHECK YOUR UNDERSTANDING

1. Describe how ameba ingest food.
2. What are vacuoles? Why does a cell need them?
3. What is diffusion?
4. Why is circulation so simple in the protists?
5. List three important things that bacteria do that are helpful to humans.

KEY IDEA #3:
Fungi are a group of organisms that belong to a fifth kingdom, separate from plants, animals, protists, and monerans.

The Fungi Kingdom

Why is it necessary to have a separate kingdom for fungi? Fungi are different from plants. Fungi do not make their own food. They have no chloroplasts or chlorophyll. Fungi do not have leaves, stems, or real roots with root hairs. Fungi are different from animals. They do not move from place to place. They are non-vascular organisms that reproduce by means of spores. Growth in fungi is by threads of hypae. The branches form a cover that is mycelial growth. Fungi are grouped alone in the fungi kingdom.

Life Activities of Fungi
Fungi grow in areas of rotting leaves and wood or decaying animal matter. Fungi are the great **decomposers**

---

of the world. They produce substances called **enzymes**. The enzymes dissolve surrounding material that the fungus absorbs or takes in as food. Fungi are important in the cycle that involves breaking down waste matter such as dead leaves, rotting wood, and decaying animal life. The dissolved chemicals are useful to other organisms. Trees can take in the chemicals from the soil around their roots.

The main growth of fungi is a mass of interconnected tubes. The greater part of a fungus is below the surface on which it is growing. What you see above the surface is the spore-producing part. Fungi do not grow tall or make any woody tissue.

Molds

Most people are familiar with **mold**. Bread gets moldy, and so does jelly, fruit, meat, and cheese. Even in the cold of the refrigerator things get moldy. Mold grows best in moist and dark places. Mold grows on different materials, such as wood, cloth, leather, paper, and on many living things.

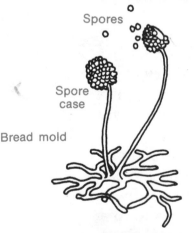

Spores

Spore case

Bread mold

Bread Mold.

The spores of the mold *Rhizopus* are in the air almost everywhere. Set a slice of moist bread on a saucer, and spores will land on it. As the spores grow, they send threads of mold into the bread. The mold makes digestive enzymes that dissolve the "food" in the bread for its own use. Some of the bread you buy has preservatives in it to keep the mold away.

Yeast

Yeast qualifies as a fungus because it lacks chlorophyll. It also uses spores for reproduction the way that other fungi do. Yeast also reproduces by budding. Yeast spores are everywhere. They are carried by air currents.

Yeast is used in baking to make the bread rise. Yeast cells cause fermentation in carbohydrates. Enzymes from yeast break down the carbohydrates into carbon dioxide and alcohol.

Yeast cells

In grapes the carbohydrates are sugar. When yeast cells mix with the sugar, the juice from the grapes becomes wine. Beer is produced with the help of yeast cells. German brewers brought special yeast cultures with them when they came to America. They continued the German tradition of beer making in this country.

Yeast cells are often studied in biology classes because they are easy to grow in sugar water. Their method of forming buds that grow into new cells is easy to see under the microscope. Learning how a population of yeast cells lives, grows, and dies is a part of the study of biology.

Mushrooms

The best known of the fungi are the mushrooms. Mushrooms grow in damp and dark places like a cave or a forest floor. Mushrooms come in many colors. Although white and brown are the most common, you can also find purple, red, and yellow mushrooms. What you see and what people eat is the fruiting body. Spores are made in this umbrella-shaped part. Most of the mushroom is found beneath the surface. Underground is a mass of what looks like white threads all strung out.

Mushrooms

Rusts and Smuts

Rusts and smuts are fungi that are parasites that live on crop plants. They are destructive parasites because they can kill the host plant. A host plant is one on which parasites feed and grow. These parasites cause millions of dollars of damage to crop plants every year.

Rusts are reddish brown fungi that grow on plants like wheat, corn, barley, and rye. The plant turns the color of rust.

Smuts are fungi that grow on cereal crops. Smut is noted for its foul smell. This fungus changes plant organs into dark masses of spores. It is most familiar on corn plants.

Penicillium

Penicillin

Some molds are not pests at all. They are helpful. The genus *Penicillium* is the mold that is effective in attacking and reducing bacteria. Alexander Fleming discovered penicillin. Quite by chance he observed that the mold was killing off bacteria in the material that he had been working with. This discovery led to the use of penicillin as a medicine for killing microorganisms that are harmful to humans.

Molds and Cheeses

The flavor of some kinds of cheeses is caused by species of molds. *Penicillium roquefortii* is the name of the mold that provides the flavor for Roquefort cheese.

Some Famous Fungi

In addition to penicillin, another famous fungus is the one that infected the potato crop in Ireland in the middle of the 1800s. This blight fungus caused a serious famine in Ireland. Many people died of starvation. As a result of the blight on the potato crop, many people left Ireland and came to settle in the United States.

Another famous fungus is the one that causes "athlete's foot." This fungus infection of the feet is widespread. It causes discomfort, but it is not deadly. Fungi that can be deadly, however, are wild mushrooms. They can poison people.

Ringworm is a fungus infection of the human scalp and skin. It grows in ring patterns. The "worm" part of "ringworm" is another example of how common names can be confusing; the cause of ringworm is not a worm at all. It is caused by a fungus.

Lichens

Lichens are unusual because they are really two organisms growing together. Lichens are fungi and algae living together in a partnership. This partnership is called **mutualism** or **symbiosis** because the fungi and the algae depend upon each other. Just as in most partnerships, each partner has a job to do. In the lichen partnership the fungus provides moisture because of its ability to absorb and hold water. The alga has chlorophyll and is able to make food that the fungus can use.

Lichens are food for animals. One famous lichen is the "reindeer moss" which provides food for caribou herds, a kind of northern reindeer.

Lichens can grow in places where other organisms cannot survive. For this reason they are often the first to grow in a new area, and are sometimes called "pioneer" organisms. Lichens are able to withstand great extremes of temperature and are often found in very cold places, such as near the North and South Poles.

Lichens can live nearly everywhere. They live close to the surface where they attach. Lichens can be found on the barks of trees, on the surface of bare rocks, and on barren mountain tops. Different species with a variety of colors often "paint" rocks into striking natural patterns.

Lichens are very sensitive to pollution and are not usually found in cities.

CHECK YOUR UNDERSTANDING

1. What characteristics make it necessary for fungi to be in a kingdom of their own?
2. Give the name of one of the most common fungi.
3. What are three useful things that yeasts do?
4. Where does penicillin come from? What does it do?
5. Name three famous fungi. Why are they famous?

KEY IDEA #4:
Fungi carry out basic life activities and affect the lives of other organisms.

Fungus Life Activities

How Fungi Get Food

Fungi get their food by absorption. Whatever they are living on or in is the source of food for fungi. Spores that land on bread send runners down into the bread. The runners give off digestive enzymes. These chemicals dissolve the bread. The chemicals of the bread become the energy source which the fungi use to live, grow, and reproduce.

Because fungi lack chlorophyll, they must depend entirely on either green plants or on animals that have green plants at the beginning of their food chain. Bracket fungi grow on the sides of trees. They look like little house roofs. They depend on the tree for their food and energy. Mildews grow on old leather, paper, cloth, and anything else that is a product of living things. Mildews depend on moisture and their digestive enzymes to break down the chemicals in these products for use as food.

Fungi and Roots

Some fungi live in a close relationship with the roots of green plants. The fungi take water from the soil and hold it for use by the plant. Minerals are also absorbed by the fungi. The green plant provides the fungi with nutrients which it dissolves with its digestive enzymes. There are some forest trees that do not grow as well as they might if the fungi are absent from the soil around their roots.

The fungi and green plant relationship is of mutual benefit. This is called mutualism, a special kind of living together where both members benefit.

The Great Decomposers

Fungi are the world's great decomposers. A decomposer is a microorganism that breaks down dead tissue and returns the nutrients to the soil. Fungi secrete digestive juices into the material that they are living on — leaves, soil, rotting logs, animal remains, and fruit. They digest the material for their own use, but return most of it to the soil to be reused by other organisms.

Fungi decompose all kinds of waste products, remains of plants, and remains of animals. They help recycle useful chemicals and substances.

Reproduction in the Fungi

Fungi reproduce asexually by means of spores. Spores are reproductive cells that are very small and are produced in tremendous numbers by each fungus.

Bread Mold

Bread mold reproduces by two methods. One method is asexually when a single spore germinates, or begins the growth pattern of a new cycle that ends with the making of new spores. The second method is sexually, when gametes, or sex cells, join together from two strains of mold cells. The body that is formed by sexual reproduction germinates and produces a new set of spores. Bread mold reproduces rapidly. A whole loaf of bread that the mold lands on can get covered overnight!

Growth of a Fungus

A fungus grows in moist, dark, and warm places. Growth is the result of a set of tube-like threads that mesh in with soil, bread, leaves, or fruits of all kinds. Wherever there are carbohydrates, the fungi grow. They form a mass of mostly white and often fuzzy-looking threads. Inside of the threads are many nuclei, but no cell walls. Materials for growth are transported in this fragile and moist mass of threads.

Over 100,000 Species

Specialists in the study of fungi have identified over one hundred thousand species. Fungi grow everywhere. Lichens grow in both the coldest and the hottest places on earth. Their spores are in the air you breathe, in the waters and the soils of the world. You already know two that live on humans—the fungus of athlete's foot and the fungus that causes ringworm.

CHECK YOUR UNDERSTANDING
1. How do the fungi get their food?
2. What is mutualism?
3. Why are the fungi called the "great decomposers?"
4. Describe reproduction in the fungi.
5. Where are some of the places that the fungi grow?

Summary of Chapter 5

Protists, monerans, and fungi are so different from plants and animals that they are classified separately into their own kingdoms.

Protists are simple animal-like protozoans and plant-like organisms called algae. Protists play important parts in the lives of other organisms. Chlorophyll in the protist, green algae, is important in the food-making process. Protists reproduce asexually. Many of the protists use mitosis and cell division to produce the next generation. Protists do all of the things that other organisms do.

The kingdom monera is the kingdom of the bacteria. They are the most numerous organisms on earth. There are many useful bacteria that help in food processing. Bacteria can be harmful by causing disease and spoiling food. The most important thing that bacteria do is to fix the nitrogen of the air into useful chemicals for plants to use in making proteins.

Fungi are different because they lack chlorophyll and live on other organisms or materials. Fungi are especially important as decomposers. They are responsible for decomposing, or breaking down, dead animals and plants and putting this changed matter back into the soil, so it can be re-used by other organisms. Fungi reproduce asexually. Some fungi produce different strains that form gametes and reproduce by sexual means.

1. fungi
2. algae
3. spirogyra
4. chloroplast
5. chlorophyll
6. protococcus
7. brown algae
8. diatoms
9. cilia
10. sporozoans
11. euglena
12. gullet
13. vacuole
14. stimuli
15. asexually
16. daughter cells
17. mitosis
18. gamete
19. decomposer
20. enzymes
21. budding
22. molds
23. rusts
24. smuts
25. mutualism
26. mushrooms
27. yeasts
28. fermentation
29. penicillin
30. parasites
31. blight
32. ringworm
33. lichens
34. absorption
35. mildews

1. What characteristics make a separate kingdom for protists?

2. What part do the plant-like protists of the ocean play in the amount of oxygen in the air?

3. What is the main difference between spirogyra and protococcus?

4. What is the connection between sea lettuce and regular lettuce?

5. What organism has a shell made of the same material used to make glass?

6. What organism is a ciliate? Tell how it moves.

7. What is the cause of the disease known as malaria?

8. Why is the *Euglena* so hard to put in one group?

9. What are two ways that the animal-like protists obtain their food?

10. Tell how diffusion works in the paramecium.

11. What is the purpose of the euglena's eye spot?

12. What is mitosis?

13. What are the two ways that diatoms reproduce?

14. What are two ways in which bacteria is useful to people?

15. What do the enzymes of the fungi do?

16. What is fermentation?

17. What is the relationship between a parasite and a host plant?

18. In the lichen partnership, how do the algae and the fungi help each other?

19. Where do mildews grow? What do they use for food?

20. Why are fungi called the great decomposers?

To discover how animals use their body systems to carry out basic life activities in order to stay alive.

KEY IDEAS

Animals use a variety of food-getting and food-ingesting methods to maintain their bodies.

Digesting food and eliminating or getting rid of waste products are necessary to help animals stay alive.

Systems for circulation and respiration provide food and oxygen to all animals' cells.

Coordination of animal activities is done by a nerve network or nervous system.

CHAPTER 6

Discovering How Animals Stay Alive

Introduction

Supporting and maintaining the cells, tissues, organs, and organ systems is the key to staying alive. Some of the activities that support life are getting food, moving, growing, using food, and removing waste.

KEY IDEA #1:
Animals use a variety of food-getting and food-ingesting methods to maintain their bodies.

Food Getting and Ingesting

You can understand more clearly some of the ways animals get food by studying five different animals.

Methods of Getting Food in the Sponges

Sponges are the simplest multi-cellular animals. They have a system for pumping water in and out of their bodies. In the process of pumping, sponges take in diatoms and other small organisms for food. This food is their source of energy. Sponges are attached to one spot on the ocean floor. They are known in biology as **sessile** animals: animals that are attached to one spot by some kind of *holdfast cells*. Since they are not able to move around to get food, they must continually pump lots of water every hour of every day.

Sponges eat organisms of very small size. Huge numbers of organisms must be "eaten" by the sponge. The sponge holds tight to its place and pumps many gallons of water filled with tiny organisms. Food-getting in sponges is a simple plan for a simple animal. Sponges must live in quietly-moving water in order to obtain food.

Food Getting in Hydras and Jellyfish

Hydras are sessile animals for most of their life. They attach by a holdfast, stretch their bodies and tentacles to great lengths, and sting and capture small water animals. Hydras are freshwater inhabitants of ponds and streams. They are predators that use their **tentacles** to bring their **prey** into a central opening or mouth. *Prey*

A hydra ingesting its prey

organisms are animals taken by other animals as food. Food is ingested by the movement of the hydra's tentacles.

Jellyfish get their food as they float in their marine environment with their tentacles streaming along beneath them. Fish that run into the tentacles are stung and captured. The tentacles bring the food to the central part of the jellyfish, where it is ingested.

Jellyfish also take in microscopic green algae that float about in the sea. These algae are members of the protist kingdom. They are important producers of food and oxygen. There is a great variety of these algae. All together, these algae are called plankton. The plankton of the seas is the main source of food and energy for larger organisms.

How Spiders Get Food

The nest or the web is the spider's trademark. Spiders build elaborate nests and webs to capture insects. Once caught, the insect prey is wrapped and the blood is sucked from its body.

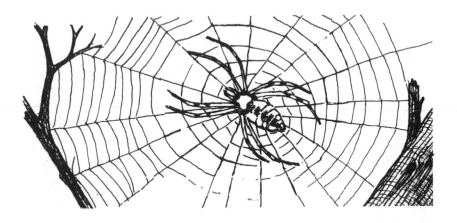

A spider uses its web to get food.

How Insects Get Food

Some insects eat plants by using cutting jaws on their mouthparts. Walking sticks chew on tree and plant leaves. Termites eat wood, and may eat away the whole foundation of a wooden building. They have chewing mouthparts. The ant lion, or doodle-bug, feeds on ants that are caught in sliding sand pits. Adult dragonflies eat insects which they capture out of the air. These beautiful insects also eat mosquito larvae.

The largest group of insects in the world are the beetles. One of them is called the confused flour beetle, and it has chewing mouthparts to eat flour, grain, and anything starchy like beans or peas. The flour beetle is often found around cereal or cornmeal. Another member of this group is the Japanese beetle, which is a pest that feeds on the leaves of grapevines, apple trees, cherry trees, corn, roses, blackberries, strawberries, and hollyhocks.

One whole group of insects are called bugs. They have sucking mouthparts and include the squash bug, bedbug, backswimmer, and water boatman.

Butterflies have sucking mouthparts to siphon the nectar from flowers. The larvae, or immature, stage of butterflies are caterpillars that chew the leaves of many plants. The periodical cicada has a sucking beak to take juices from roots and stems.

The horsefly and the deerfly have piercing mouthparts and live on the blood of warm-blooded mammals. The common housefly feeds on dead and decaying matter. The housefly has the scientific name *Musca domestica* and should not be confused with the horsefly.

How Frogs and Toads Get Food

Frogs and toads are amphibians and live much of their lives around water or in moist places. They eat insects, earthworms, and other small animals. The tongue of these animals is attached to the floor of the mouth. It lashes out with a flicking motion, and faster than the human eye can see, it captures a flying insect. The end of the frog's tongue is sticky, and the insect has little chance to escape.

Frogs and toads capture prey with their tongues.

Basic Biology

The bullfrog, *Rana catasbeiana*, is a large frog with a diet of insects, other frogs, fish, and even birds. Bullfrogs are famous for their great jumping skills. Many fine restaurants serve the legs of these frogs as a delicacy.

Another famous amphibian is the American toad, *Bufo americanus*. The toad lashes out its sticky tongue at ants, slugs, earthworms, beetles, and bugs in general. It is a frequent inhabitant of backyards.

CHECK YOUR UNDERSTANDING

1. What are the scientific words for breathing, eating, and drinking?
2. Describe three food-getting activities in animals.
3. What do sponges do to get food?
4. What is the special feature of a spider's life?
5. What is the method for getting food among the beetles?

KEY IDEA #2:

Digesting food and eliminating or getting rid of waste products are necessary for animals to stay alive.

Digestion and Excretion

Digestion is a basic life activity in which food is broken down, turned into a liquid with dissolved small particles and then into nutrients or simpler chemicals so that the nutrients can be used. Special chemicals called enzymes mix with and dissolve food and causes breakdown into smaller chemicals.

Excretion is a basic life activity in which waste matter is removed from cells or from animal bodies. Liquid wastes are removed by some form of the excretory system. The digestive system removes solid wastes.

Digestion and Excretion in Sponges

Even simple organisms like the sponges have enzymes in their body cells. The enzymes are chemicals that act on food that came in along with the water being pumped through the body wall of the sponge. The body wall of the sponges is one cell layer thick, and yet the whole process of digestion takes place there. Enzyme action digests the food, and cells use the energy to keep the animal alive.

Energy is released from the food and is used to carry out the basic life activities. Carbon dioxide and other waste products are formed. The sponge gives off the waste materials through cells in the body wall. The simple organization of the sponge body makes digestion and excretion direct and simple processes.

Digestion and Excretion in Hydras

Hydras have tentacles that are used to capture food in the form of small water animals. The food is brought to the mouth opening and into a hollow body sac. This is where **digestion** takes place. The cells lining this sac give off digestive juices called enzymes.

The foodstuff is reduced to usable small chemicals that go to the cells. Cells use the food for energy and release waste products into the body sac. Water and carbon dioxide are the main waste materials formed from cell respiration. They are removed through the mouth opening.

Digestion in Spiders

Digestion in spiders is unusual. It takes place outside of the body of the spider for the most part. Spiders feed on insects. They feed on the soft inner parts. To do this, spiders use a pair of special fangs near the mouth. The fangs first inject poison to kill the prey. Then holes are punched to allow digestive juices from the spider's mouth to enter the prey's body.

The digestive juices turn the soft inside parts of the prey into a fluid. Spiders take in only liquid food. The fluid is sucked up into the spider's body and into its stomach where further digestion takes place. Food molecules are absorbed by the blood.

Digestion in Insects

One example of an insect is the grasshopper. Grasshoppers eat grass and have mouth parts that chew. The digestive system of the grasshopper is a tube that begins with the mouth. The mouth is connected to the esophagus, which in turn, is connected to the crop. The crop is a storage place and leads to the **gizzard**, where food is ground up. The ground-up food goes through to the midgut where enzymes digest the food. The food molecules are absorbed through the stomach wall into the bloodstream. After digestion, some solid wastes move from the midgut through the hindgut and out of the grasshopper through the **anus**.

Digestion and Excretion in Frogs and Toads

Frogs and toads are great insect-eaters. They are valued by humans because they help control the populations of insects. They have large, wide mouths that help in the capture of insects.

The digestive system of these two amphibians is similar to that of the vertebrates in general. In fact, frogs are studied in some detail in biology courses because they can be compared to the human. They have similar organs of digestion, similar kinds of digestive glands, and a similar blood system.

Food that enters the mouth is pushed into an opening at the rear of the mouth called the gullet. Then the food moves to the esophagus, which is a tube that connects to the stomach. Not much happens to the food until it reaches the stomach, where gastric juices in the form of enzymes begin to break the food apart. The next stop is the small intestine, where bile from the liver and digestive enzymes from the pancreas further break down the food. Frogs and toads and other vertebrates have a gall bladder that stores and releases the bile that breaks down fats into molecules of smaller size.

The food mass that is formed by digestion is now a liquid in a usable chemical form. The **blood vessles** that surround and are imbedded in the small intestine wall absorb the food material and circulate it to all parts of the body. The cells use the food for energy to move muscles and to carry out basic life activities.

There are some waste materials that are left after digestion of the main part of the food. These semi-solid products are moved on to the large intestine, and from there to the **cloaca**, and out of the body through a small opening called the anus.

Meanwhile, back in the body cavity, the kidneys are filtering waste products from cell respiration from the blood. This liquid waste is called urine, and is stored temporarily in the **bladder** before being released to the cloaca.

The gaseous waste product of cell respiration is carbon dioxide, and it is released and eliminated through the skin, lungs, and mouth.

The digestive system of the vertebrates is also called the alimentary canal. It is a system of organs that maintain the life of the organism. It is closely associated with the excretory system and with the circulatory system.

1. Tell about digestion.

2. What is excretion?

3. How do hydras digest their food and dispose of wastes?

4. What makes digestion in the spiders so unusual?

5. What happens to food in the small intestine of the frog?

KEY IDEA #3:

Systems for circulation and respiration provide food and oxygen to all animals' cells.

Circulation and Respiration

Organisms must get food, oxygen, and water to their body cells. Animals use a pump and a circulatory system of conducting tubes, or vessels, to do that job. The pump is the heart, and the vessels are arteries and veins. The circulatory system is a transport system.

Organisms must get oxygen to their body cells. A breathing, or respiratory, system of some kind does that job. Vertebrate animals rely on a system of gills, windpipes, lungs, and blood vessels to get oxygen to all the right places.

Circulation and Respiration in Hydras and Jellyfish

The cells of hydras are in direct contact with the water environment. Circulation of water carries materials into and out of the cells. Inside the hydra's body, food materials are digested and circulated directly into the cells that line the inside wall. Waste materials that result from cell respiration—water and carbon dioxide — diffuse from the cells to the body cavity and are expelled through the mouth opening or through the body wall. The body plans for circulation and respiration in the hydras and jellyfish are simple compared to the vertebrates.

Circulation and Respiration in Insects

Insects are very active organisms which require energy. Their circulatory system is an open one. There is a heart in the upper part of the body cavity. It is not a complicated heart, just a swelling of a tube called the *dorsal aorta*.

Insect blood is a watery fluid. It moves food, oxygen, and carbon dioxide around the body parts. Insects have a special feature in their respiratory system: tubes connect the outside air with the inside body. The openings of the small tubes run along the sides of the insect body and open and close in rhythm. The abdomen moves somewhat like an accordion. Air is forced in and out quite rapidly to take care of the needs of the active insect.

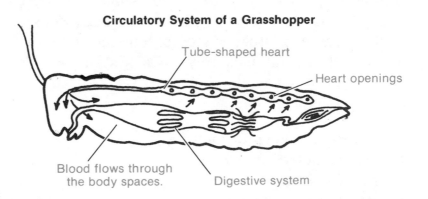

Circulatory System of a Grasshopper

Tube-shaped heart

Heart openings

Blood flows through the body spaces.

Digestive system

Circulation and Respiration in Spiders

Spiders have a set of "book" lungs located in the body cavity. This is in addition to their air tubes. The **book lungs** get their name from their appearance of pages in

Basic Biology

a book with spaces between them. The air spaces are connected to the outside air through an opening called a spiracle. Blood circulates in the "pages," and plenty of air is in contact with the blood. Exchange of oxygen takes place. The oxygen is used in cell respiration in the spider's cells.

Circulation and Respiration in Amphibians

The circulatory system in the amphibians is much more complex and highly organized than in the sponges, hydras, jellyfish, insects, and spiders. The problems are bigger. In frogs there is much more cell activity deep inside the body. There are many more layers of cells. The cells are not directly exposed in the watery environment as are those of the sponges, hydras, and jellyfish. The body plan of frogs and other vertebrate animals must be able to circulate materials to a much deeper and more spread-out variety of cells.

The frog heart has three chambers. The chambers are connected to each other and to blood vessels. Blood circulates in vessels to carry oxygen to body cells. Blood picks up and gets rid of carbon dioxide. Blood vessels carry blood from the heart that is full of oxygen to the head, legs, and main body. Another set of blood vessels returns blood to the heart that is full of waste carbon dioxide from the body, legs, and head. The heart is the manager of what blood goes where. The chambers serve as collectors, sorters, and senders.

Within the body of frogs and other vertebrates there are two kinds of circulation. One kind works on circulation to and from the lungs—**lung circulation**. The other works on circulation to and from the body—**body circulation**. Both circulations use blood vessels to transport the blood. The heart is the pumping organ that is in the middle of the two different circulations lung circulation and body circulation.

Each circulation forms a complete circuit. The lung circuit is from heart to lungs and from lungs to heart.

The body circuit is from heart to body parts and back to the heart. Arteries are the blood vessels that take the blood away from the heart. Veins are the blood vessels that return the blood to the heart. Artery blood is full of oxygen. Arteries get their oxygen supply in the lungs and through the skin of the frog. The blood takes the carbon dioxide out of the frog by way of the lungs.

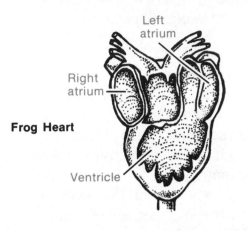

Frog Heart

Left atrium

Right atrium

Ventricle

The purpose of the circulation of blood is to get oxygen to all parts of the body. The frog's cells use the oxygen to carry out cell respiration. Food taken in by the frog goes through the digestive system and ends up in the small intestine. Blood vessels along the intestine take up the dissolved food and transport it to the cells, where energy is released.

Frogs spend much of their time in water. They have a thin skin. They are able to take oxygen through their skin to their body cells and blood vessels by diffusion. They also get rid of some waste carbon dioxide and water through their skin.

1. What is the function of a circulatory system?
2. What does the respiratory system do?
3. What is a special feature of the respiratory system in the insects?
4. What method do spiders use for getting oxygen?
5. Name two kinds of circulation in frogs.

KEY IDEA #4:
Coordination of animal activities is done by a nerve network or by a nervous system.

Coordination

Coordination of body activities means getting the different parts to act smoothly together. Coordination of body activities in the invertebrates, like hydras, is carried out by simple nerve networks. These nerve networks respond to objects and changes in their surroundings.

Coordination in multi-cell vertebrate animals is not so simple. Frogs, birds, raccoons, horses, and humans have complicated body systems. The systems must all work well together to get things done. Muscles, organs, and bones must work smoothly for these animals to stay alive. Coordinating the work of the muscles, organs, and bones is the job of the **nervous system**.

Coordination in the Sponges

The sponges are simple animals that have no centralized or highly-developed nervous system. Sponges do not have complicated problems to solve. They dwell in warm water where conditions stay much the same.

The cells of sponges carry out simple activities. Each kind of cell does a different job. Cells in sponges act alone; they do not form into tissues. Separate parts of sponges are so independent that if a sponge is cut into pieces, each piece is able to re-grow into a new sponge.

Coordination in Hydras and Jellyfish

Unlike the sponges, the more highly developed hydras and jellyfish have tentacles that must be coordinated. Hydras are attached and move their tentacles about, but they can also move about independantly. Hydras have been observed doing "tentacles over holdfasts" similar to human gymnasts doing handsprings. This is possible because of communication between nerve cells and myofibrils that act like muscle cells.

A hydra can move by flipping over.

Hydras react to the touch of an object in their environment by pulling all body parts together. This happens because of a **nerve net** that is a kind of simple nervous system. This nerve net coordinates the movement of the hydra's tentacles as it captures food and then brings the food into an opening or "mouth" at the base of the tentacles. If chemicals are put into the water, or if the temperature is altered, the body of the hydra responds by pulling its parts together.

Jellyfish pulsate and move around by means of a muscle-like ring of cells. The tentacles bring captured food into the center of the body. Coordinated food-getting and movement are done by a nerve net similar to that of the hydra.

Coordination in the Insects

Insects have sense organs. Eyes, antennas, and a **tympanic membrane,** or eardrum, are all used to respond to stimuli from the environment. Reach for a grasshopper and it quickly jumps or flies away. The bee diagram below shows a brain, a nerve cord running the length of the body, and a pair of compound eyes. Most insects have these kinds of parts in their nervous systems.

Insects eat a variety of foods, use a variety of methods for getting food, and have a wide variety of behavior and movement patterns. All of this requires a high degree of coordination. Insects are among the most successful of all kinds of organisms on earth. Part of their success can be credited to their coordinating system. The brain of most insects has a highly developed set of **optic lobes**. This part of the brain coordinates eye movements. Muscles in the legs, in the area of mouth parts, and at the base of the wings act together in food-getting or escaping from being food for some other organism.

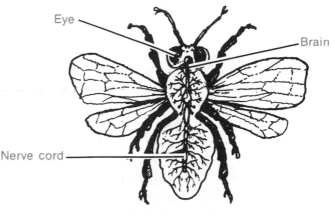

Nervous System of Honey Bee

Coordination in Spiders

Spiders have a ventral nerve cord that attaches to a fairly large brain that enables them to carry on a variety of activities. Spiders that weave webs do so by instinct. They do not have to be trained, or learn how. Web-making is an inherited or inborn skill.

To make their webs and wrap their prey, spiders make silk. On the underside of the spider are **spinnerets**, organs that make silk threads. The spider is a web-builder and a predator. These activities require coordination.

Tarantula

Spiders have eight simple eyes that are connected by nervous tissues to the brain. These eyes can detect light but cannot form images. The presence of an insect trapped in the web is sensed by sensory hairs on the legs of the spider. The tarantula is covered with velvety hairs and relies on its sensory touch.

Coordination of Frog Activities

The nervous system of the frog is like that of most vertebrate animals. Biology classes study the frog's anatomy to understand how human systems look and work.

All of the activities of a frog's body are coordinated by the nervous system. The nervous system of the frog is made up of the brain, the **spinal cord**, sensory organs, and nerves. One set of nerves runs from the brain to parts of the head. The eyes are important sense organs that are connected by nerves to the brain. All actions of the nervous system are very fast. That is a feature of nervous tissue.

The Spinal Cord. A set of nerves connects the spinal cord to all parts of the body cavity, or abdomen. The nervous system acts very much like the conductor of a symphony orchestra. It controls the other systems of the body. It does this because most cells of the body have nerve endings that receive signals through the spinal column. Nerve cells go to all parts of an organism. A main feature of cells is that they are sensitive to a number of stimuli. Most of the muscles in the frog's body respond to directions from the brain. One set of nerves controls voluntary muscle actions by way of the brain.

Automatic Actions. A separate set of nerves controls all of the automatic actions of the body. The heartbeat, changes in sizes of blood vessels, and action of the digestive system are controlled by automatic actions of nervous tissues. These activities are coordinated directly by the spinal cord and brain.

Basic Biology

The Brain. The brain is the organ of the nervous system that is in charge and at work all the time. Seeing, feeling, sensing, smelling, and hearing are all **coordinated** in the brain. Even tasting gets some action in the brain.

A set of muscles that moves has received a signal from the brain on a **motor nerve.** The original signal may have started when the frog was touched. The nerve endings in the skin sent a signal on a **sensory nerve** to the brain. Motor nerves and sensory nerves are the coordinating network of vertebrate animals.

Being alive is a matter of responding to the large numbers of stimuli in the environment. Frogs hear other frogs during the mating season. They respond to the mating call of their species.

Frogs Are Complex. Unlike the sponges, hydras, jellyfish, and spiders, the frog is able to do a much greater number of things. Frogs can leap, flick out their tongues to capture insects, swim with great speed, make sounds, and escape the grasp of a human trying capture it. Frogs reproduce by sexual reproduction that is coordinated by their nervous systems.

CHECK YOUR UNDERSTANDING

1. What is the job of the nervous system?
2. What is the "nerve net" in hydras?
3. What are the main parts of the nervous system in insects?
4. What are the main parts of the nervous system in the frog?
5. Tell the difference between a motor nerve and a sensory nerve.

Summary of Chapter 6

Animals stay alive by carrying out basic life activities. Animals have different ways of ingesting food, and they use a variety of methods for digesting food. The variety in methods runs from the simple "sitting and getting" to the complex flicking of a frog's tongue to capture a flying insect. From the stinging tentacles of jellyfish to the biting jaws of the doodlebug, the methods of getting food are very different.

Digestion and excretion require systems from the simple diffusion in the sponges and jellyfish to the complicated systems of the frogs and toads. Breaking down animal and plant food into useful chemicals of food by the action of enzymes helps animals stay alive. Spiders live on a liquid food diet. Insects have a tube-like digestive system with special organs that do special jobs along the route. Getting rid of liquid waste is the job of the excretory system. Carbon dioxide waste material is also given off by animals through their respiratory system.

The circulatory and respiratory systems work together to transport important materials in animal bodies. Digested food, oxygen, and carbon dioxide are the main passengers in the fluids of the circulatory system. Lungs and air tubes help bring in oxygen and get rid of carbon dioxide. Insects and spiders with the early beginnings of real hearts pump blood through the body. Frogs and toads have three-chambered hearts that take care of lung circulation and body circulation. Frogs also take oxygen from their water environment and the air through their skin.

Coordination of body activities is done by nerve networks and by complex nervous systems. Insects have brains, nerve cords, and connecting nerves. Frogs and toads are representative vertebrate animals with brains, spinal cords, sensory and motor nerves, and organs such as eyes, ears, and specialized mouths. Spiders build complex webs by instinct or inborn patterns of nervous behavior. The nervous system in insects is responsible for a wide variety of activities—flying, jumping, swimming, crawling, fighting, mating, egg-laying, and making sounds. Staying alive means carrying out all of the basic life activities in a coordinated and consistent way.

1. *sessile*
2. *tentacles*
3. *bladder*
4. *digestion*
5. *excretion*
6. *esophagus*
7. *gizzard*
8. *anus*
9. *liver*
10. *cloaca*
11. *respiration*
12. *book lungs*
13. *lung circulation*
14. *body circulation*
15. *blood vessels*
16. *cell respiration*
17. *coordination*
18. *spinal cord*
19. *nerve net*
20. *nervous system*
21. *tympanic membrane*
22. *optic lobes*
23. *spinnerets*
24. *motor nerve*
25. *sensory nerve*

1. What are some of the essential things that animals must do to stay alive?

2. Tell the difference between ingestion and digestion.

3. How do sponges get their food?

4. Tell about some of the different kinds of mouthparts among the insects.

5. What is so special about the way that frogs and toads use their tongues?

6. What do enzymes do?

7. Name the eight parts of the grasshopper's digestive system.

8. What are the nine parts of the digestive system in a frog?

9. What is another name for the digestive system of the vertebrates?

10. What system in an organism's body is used to move or transport food and oxygen?

11. What is the main job of a respiratory system?

12. What are book lungs? What makes them work so well?

13. What does the skin of a frog have to do with oxygen and carbon dioxide?

14. What is a nerve net?

15. What is the difference between the tympanic membrane and an optic lobe?

16. What are the main parts of the nervous system in the insects?

17. What are the main parts of the nervous system of a frog?

18. What do motor nerves do?

19. What is the job of a sensory nerve?

CHAPTER GOALS:

To discover how plants live their lives by transporting food and water, by making food, and by giving off oxygen while taking in carbon dioxide.

To understand the way in which special plant parts are used for reproduction.

KEY IDEAS:

Seed plants and ferns use vascular tubes to move food, water, and minerals through their roots, stems, and leaves.

Green plants use carbon dioxide, water, energy from the sun, and the chlorophyll in their cells to make food.

Green plants release oxygen into the air during the food-making process called photosynthesis.

Plants have flowers, fruits, seeds, and spores as reproductive organs.

CHAPTER 7

Discovering How Plants Live Their Lives

Introduction

What if you couldn't move? You wouldn't be able to go to your favorite fast-food place. If you were able to get there, all you could do is to stand in one spot. You couldn't even open your mouth or swallow. You would be like a plant and would need to grow some roots, stems, and leaves. You would have a system of vascular tubes instead of blood vessels. Your flowers, fruits, and seeds would be parts of your reproductive system. Or, if you were a moss, you would reproduce by spores. Instead of breathing out carbon dioxide, you would exhale oxygen. And, instead of blood cells, you would have special green cells called **chloroplasts**. The chloroplasts would enable you to make your own food. You wouldn't even need that fast-food place,

The process of photosynthesis works like a factory. Plants make their food by photosynthesis.

because you could make the food you needed. In fact, you could feed the animals of the world. They would be dependent upon you, not only for their food, but also for their oxygen.

The Transport System in Vascular Plants

Roots Absorb Water and Minerals

The vascular system in plants begins with the roots. Roots grow under the ground. What do roots do? Roots hold the plant in one place. They take in water and minerals from the ground. They can soak up, or absorb, the water and minerals and store them until needed. Or, the roots can transport the water and minerals to other places in the plant, using their **xylem tubes**.

How Roots Work

Roots grow into and through soil. Roots are built to be able to push their way around in the ground and find what they need. They grow longer to reach deep ground waters. Roots often grow around rocks under the ground and around the roots of other plants. Have you ever tried to pull out a weed growing in your yard? Were you surprised at how strongly it held on to the ground? Were you surprised at how deep or widespread the root system

Root hairs

Root cap —

had grown in the soil? Many root systems under the ground are as large as and sometimes larger than the tree or bush that they support above the ground.

Parts of a Root

Root hairs are thin tubes that extend from the root cell wall. They are the connection with the water and minerals in the soil. There are millions and millions of such small tubes that actively soak up water and minerals from the ground water supply.

A special structure at the end of each root is the **root cap**. It protects the main part of the root as it grows and pushes its way through the soil. Root cap cells wear out as the root grows, so new root cap cells are made all the time.

Inside the root, and up into the stems and leaves, xylem tubes transport incoming water and minerals. Above the ground and down into the root cells, special tubes called **phloem tubes** carry food from the leaves to all parts of the plant.

Stems for Support

A stem is the part of a plant's body that supports the leaves and connects the leaves and the roots. Stems store food and transport food, water, and minerals through the plant.

Cross section of tree with growth rings

Plants also contain a layer of tissue that produces new xylem and phloem cells. The layers are responsible for new growth. The rings that you count to tell the age of a tree are called "annual rings." These rings, found in woody stems, are layers of tissue.

The Leaves of Plants

Most leaves are thin and flat with three main parts: the **petiole**, the blade, and the **veins**. The petiole attaches the leaf to the stem or to a branch. The blade is the flat part or surface where food is made. The veins are scattered through the blade and run through the stalk to the stem. The blade is built to receive and use light from the sun. A single leaf has a large surface area. A whole tree full of leaves, then, is able to gather huge amounts of energy from the sun.

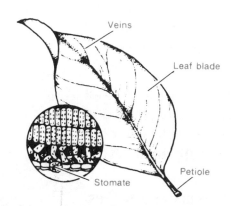

A leaf is made up of layers of cells.

The veins of leaves have a system of skinny tubes that are arranged in a pattern. Veins transport food and water between the main stems and the leaf. Small openings on the underside layer of each leaf allow gases like carbon dioxide and oxygen to enter and leave the leaf. Water vapor, an invisible gas, also enters and leaves through these openings called **stomates**, or pores. Most of the water gathered by the roots passes through the vascular system of the plant and is transferred to the air through the pores.

CHECK YOUR UNDERSTANDING

1. What function do roots have in plants?
2. Name the three main parts of a leaf.
3. Tell what stomates or pores do in a leaf.
4. What job do veins have in a leaf?

KEY IDEA #2:

Green plants use carbon dioxide, water, energy from the sun, and the chlorophyll in their cells to make food.

Food Making in Green Plants

Special cells in the plant leaf hold a green pigment called **chlorophyll**. Chlorophyll uses the energy of sunlight to change carbon dioxide and water into simple sugars and oxygen. The simple sugar, or glucose, is food for the plant. The process is called **photosynthesis.** Photosynthesis is a process plants use that takes in sunlight, carbon dioxide, and water and produces glucose, oxygen, and water.

How Photosynthesis Works

Plant leaves contain special cells called chloroplasts. Chloroplasts contain chlorophyll, a green substance. When sunlight hits a leaf, the chlorophyll captures the light energy and keeps it there. Carbon dioxide comes from the air and enters the leaf through the stomates. Water comes up through the roots. Water also enters through the stomates.

The plant uses the energy from the sun to split up the water into its two parts: hydrogen and oxygen. The oxygen from the water is released into the air. Hydrogen, the other element found in water, is combined with carbon dioxide to produce simple sugars. Because of its chlorophyll, the plant is able to change **light energy** from the sun into **chemical energy**.

Chemical Energy from Food

What does chemical energy have to do with making the food that you eat? Simple sugars are carbohydrates. Sugar is food. The food you eat contains carbohydrates, fats, and proteins.

The process of photosynthesis is the food-making process in plants. The simple sugars made during photosynthesis combine with other molecules to make different kinds of carbohydrates, such as starches. They also combine with other molecules to form fats.

How Chemicals Combine in Photosynthesis

During photosynthesis, the energy given off by the chlorophyll breaks up water molecules into hydrogen and oxygen. Hydrogen combines with carbon dioxide to form a simple sugar called glucose. The breaking up and the combining of molecules is called a chemical reaction. Everything in the world is made up of **molecules**, and molecules are made of tiny atoms. An atom is the smallest part of a molecule. A molecule is the smallest combination of atoms which forms a separate substance. When atoms combine and recombine, different molecules are formed.

Plants use oxygen and carbon dioxide in photosynthesis and respiration.

Chemical Formulas

The chemical formula for water is H_2O. Water is made up of two parts hydrogen and one part oxygen. The H stands for the hydrogen atom, and the O is the symbol for the oxygen atom. The small number 2 is right after the symbol for hydrogen. That means that there are two hydrogen atoms in a molecule of water. There is no number after the letter O. This means that there is only one oxygen atom in a molecule of water.

For 12 molecules of water, the formula for water would look like this: $12H_2O$. How many hydrogen and oxygen atoms are in 12 molecules of water? If you said 24 hydrogen atoms and 12 oxygen atoms, you would be correct. Each molecule of water contains two hydrogen atoms (H_2). Therefore, 12 molecules of water have 12×2, or 24 atoms of hydrogen. Each water molecule has only one oxygen atom in the formula, so 12 water molecules contain 12×1, or 12 oxygen atoms.

The chemical formula for carbon dioxide is CO_2. Each molecule of carbon dioxide has one atom of carbon and two atoms of oxygen. The formula for six molecules of carbon dioxide is $6CO_2$. Six molecules of carbon dioxide contain six carbon atoms and 12 oxygen atoms.

The Formula for Photosynthesis

In the process of photosynthesis, twelve molecules of water and six molecules of carbon dioxide combine to form one molecule of sugar, six molecules of oxygen, and six molecules of water. The chemical equation for the entire process of photosynthesis looks like this:

$$12H_2O + 6CO_2 + \text{light energy} + \text{chlorophyll} = C_6H_{12}O_6 + 6O_2 + 6H_2O$$

[water + carbon dioxide + light energy + chlorophyll = simple sugar + oxygen + water]

In an equation, the left side and the right side are equal. In the equation for photosynthesis, there are an equal number of oxygen, hydrogen, and carbon atoms on each side.

1. Why are plants so important to animals?
2. What is the name of the special material inside plant leaves that enables them to carry out the process of photosynthesis? What does this material do?
3. What kind of food does a plant make?
4. Name the four items that are necessary in order to have photosynthesis.
5. List three items that are produced as a result of photosynthesis.

KEY IDEA #3:
Green plants release oxygen into the air during the food-making process called photosynthesis.

Plants Give Off Oxygen

During the food-making process in the leaves of plants, oxygen is released. Oxygen is a gas that is useful and necessary to living things.

What Is a Gas?

A gas is something that you can't see or feel, but it is really there. Do you know why it is so light that you can't feel it and so clear that you can't see it? It's because the molecules in a gas are spread out and move around quickly. In a solid, such as your desk, the molecules are packed tightly together and hardly move at all. That's why the desk is so hard.

Why You Need Oxygen

The air you breathe is composed of **carbon dioxide**, **nitrogen**, and **oxygen** gases. Oxygen is one of the key gases in the air. Oxygen is very important in the lives of plants and animals. If oxygen is absent, energy is not released from food. Oxygen unlocks food energy which is stored in the cells of plant and animal bodies. This process is called **cell respiration.**

One kind of respiration is breathing in and out. You inhale air full of oxygen and exhale air full of carbon dioxide and water. Body cells use oxygen to break apart the sugar molecules that were built in plant life. Cell respiration is a special low temperature kind of burning. Animals and plant cells have a special ability to "burn" food. You get energy when your cells burn up food. This is cell respiration.

How Plants Produce Oxygen

During the process of photosynthesis, plants use the chlorophyll in their special cells called chloroplasts. Using the chlorophyll and the energy from the sun, plants combine water and carbon dioxide to make simple sugar. Each time a plant makes one molecule of simple sugar, it also makes six molecules of oxygen and six molecules of water. The plant uses some of the oxygen for its own energy. The rest of the oxygen goes out the stomates into the air. The oxygen that is released to the air came from the water when it was broken down in photosynthesis.

Plants Release Oxygen

Plants release the oxygen they don't use into the air. The oxygen exits through the many pores in the leaf. These pores, as you know, are called stomates. Each stomate has two special cells called **guard cells**. Guard cells open and close the stomate to let water and gases in and out. Oxygen exits from the plant when the

Guard cells

Chloroplasts

Stomates on Underside of Leaf

guard cells bend apart. The space or opening between the guard cells is the pore that lets out the oxygen. During certain periods of the day, the two guard cells move back together to close the stomate, and gases are not exchanged. The opening and closing of stomates is controlled by the amount of carbon dioxide inside the guard cells. In addition to the flow of CO_2, the leaf must have 100% humidity or water content inside at all times.

Plants take in carbon dioxide and give off oxygen during photosynthesis. Animals breathe in oxygen and breathe out carbon dioxide and water during respiration. This continuous process or cycle is called the oxygen-carbon dioxide cycle.

Besides green plants, the protists of the oceans go through the same oxygen-carbon dioxide cycle. Protists supply more than half of the oxygen in the world's atmosphere.

The Oxygen-Carbon Dioxide Cycle

CHECK YOUR UNDERSTANDING

1. What is the difference between the molecules in a gas and a solid?
2. Tell about the process of cell respiration.
3. How does a plant produce extra oxygen?
4. Why do leaves need guard cells?
5. Describe the oxygen-carbon dioxide cycle.

KEY IDEA #4:

Plants have flowers, fruit, seeds, and spores as reproductive organs.

Basic Biology

Reproducing New Plants

Plants can reproduce by **sexual reproduction** or by **asexual reproduction.** The prefix "a" in the word *asexual* means "without," so the word *asexual* means "not sexual." Sexual reproduction involves two parents. One parent provides the egg cell, and the other parent provides the sperm cell. In the plant world, seed plants reproduce by sexual reproduction. In asexual reproduction, only one parent is involved. Plants with spores, such as mosses and ferns, reproduce both sexually and asexually.

Asexual Reproduction

Plant-like protists reproduce asexually. One-celled organisms divide into two cells, and each cell becomes a new organism. Simple green algae reproduce this way. Plants such as mosses and ferns reproduce by forming spores. A spore is a reproductive cell with a thick protective coating. New cells are formed when the parent cell divides. The wind can blow the spores or they can fall to new surroundings where they will develop into new plants. Moss spores can develop into male and female plants. The plants produce sperm and eggs. The sperm and eggs then reproduce sexually.

Reproduction in Mosses and Ferns

Mosses and ferns reproduce both sexually and asexually. During sexual reproduction a **sperm cell** and an **egg cell** join together to form a **zygote cell**. The sperm cell is the male cell responsible for reproduction. The egg cell is the female cell involved in reproduction. The zygote cell formed from the egg and the sperm is a new organism. It is the beginning of a new plant. A sperm cell swims to the egg cell through the moisture collected on the plant and fertilizes the egg. This is called sexual reproduction, because two different sex cells or **gametes** take part in the process.

Reproduction in Seed Plants

Seed plants reproduce sexually. The egg cells are encased in a structure called the ovary, and the sperm cells are found in **pollen grains**. For reproduction to take place, the pollen must fertilize the egg. Insects, gravity, wind, or rain can transport the pollen to the egg.

How Flowering Plants Reproduce

The flower is the special part of a plant's body which makes both egg cells and sperm cells. Flowering plants, or angiosperms, have many different kinds of flowers such as roses, tulips, and daisies. The flowers may look different, but they all contain the same kinds of reproductive parts.

In a flowering plant, or angiosperm, the **stamens** are the male organs of reproduction, and the **pistils** are the female organs. The stamens and pistils are both a part of the flower.

Flower Parts of a Tulip

The first step in reproduction is **pollination.** Pollen is transferred from the stamen to a sticky substance at the top of the pistil. If pollination takes place within one flower, it is called self-pollination. If a flower from one plant pollinates the flower from another plant, it is called cross-pollination.

Fertilization

After pollination, the pollen grain grows a tube. The pollen tube grows long enough to reach down to the ovary, which is at the bottom of the pistil. The ovary contains the eggs. When the pollen inside the pollen tube meet an egg, **fertilization** takes place. Fertilization is the beginning of reproduction. After fertilization, the ovary grows and becomes a fruit with seeds inside. The job of the fruit is to protect the seeds of the next generation of plants.

Seeds

Seeds are capable of germination to produce a new plant. Seeds contain the **embryo,** or beginning stages of a new plant. When temperature and moisture conditions are right, a seed begins to germinate. **Germination** is the beginning of development. Seeds carry all the basic information to make a new plant like the ones from which it came.

Water is the key factor in the beginning of germination. Once water has been soaked into the seed from the environment, many chemical changes begin to take place. Special chemicals called enzymes cause these changes.

All activities of living things depend upon energy. Energy comes from the stored food that is the part of every seed. This food is starter food. As the plant grows and develops leaves it can make its own food. As the plant grows, it develops all of the regular parts that you recognize — roots, stems, flowers and leaves. Later flowers, fruits, and a new set of seeds develop inside the fruit. The new seeds are both the end of one cycle of germination, growth, and development, and the beginning of the cycle that will become the next generation.

How Conifers Reproduce

A pollen grain and an egg are the reproductive cells of gymnosperms, or non-flowering plants. The cone of a gymnosperm, like the flower of an angiosperm, contains the reproductive organs of the plant. However, there is a big difference between reproduction in conifers and in flowering plants. In flowering plants, both male and female reproductive parts are most often in the same flower. In plants with cones, some cones are male, and other cones are female. Some species have trees that have separate sexes. Male ginkgo trees and female ginkgo trees are one example.

First, the pollen grains are released from the male cone. Millions of grains are scattered about and can be carried by the wind or rain. Only a few grains reach a female cone. When they do, the pollen grain grows a tube that reaches into the ovary and the egg. When the pollen and egg meet, fertilization takes place. Just as in flowering plants, pollination can take place within the same tree. This is self-pollination. Cross-pollination can occur if two trees are involved.

CHECK YOUR UNDERSTANDING

1. What is the difference between sexual and asexual reproduction?
2. What are gametes?
3. Describe the process of fertilization in a flowering plant.
4. What is the difference between self-pollination and cross-pollination?
5. Describe the difference between the organization of reproductive parts in angiosperms and gymnosperms.

Summary of Chapter 7

Plants live their lives by using their body parts to survive and to reproduce. Roots and stems give plants their support. Root hairs absorb water and minerals from the ground. In vascular plants, such as ferns and seed plants, and xylem and phloem tubes carry these substances throughout the plant.

Chlorophyll molecules use sunlight. Tiny holes in a leaf, called stomates, allow materials to flow in and out. The veins in leaves have a system of tubes that transport food and water. The pattern of the veins and the shape of the leaves help identify the plant.

Plant leaves contain a green substance called chlorophyll, which enables the plant to make its own food. Food making, or photosynthesis, depends upon light energy from the sun. Plants use chlorophyll to break up water molecules into hydrogen and oxygen. The plant then combines the hydrogen with carbon dioxide from the air to produce a simple sugar. In this chemical reaction, plants change light energy from the sun to chemical energy. Without plants, there would be no food for animals.

During photosynthesis, plants produce oxygen and water in addition to simple sugar. Plants release oxygen they don't need back into the air through the stomates in their leaves. Guard cells open and close the pores to allow materials to enter and exit. Plants take in carbon dioxide and give off oxygen. Animals inhale oxygen and exhale carbon dioxide. This exchange of gases is called the oxygen-carbon dioxide cycle.

Mosses and ferns reproduce both sexually and asexually. Spores, which are the reproductive units in these plants, can divide in half to produce new organisms. This process is called asexual reproduction, because only one parent is involved. Mosses and ferns also reproduce sexually. The two gametes, the sperm cell and the egg cell, join together to produce a zygote.

Seed plants reproduce sexually. In angiosperms, the flower contains both the male sex organ, the stamen, and the female sex organ, the pistil. First, pollen from the stamen is transferred to a sticky substance at the top of the pistil. This process is called *self-pollination,* if the male and female sex organs are in the same flower, or *cross-pollination* if two plants are involved. Next, the pollen grain grows a tube which allows the pollen grain to travel down to fertilize an egg in the ovary at the bottom of the pistil. The zygote grows and becomes a fruit with a seed inside.

Gymnosperms have cones which contain the reproductive organs. Pollen grains of male cones are scattered about by the wind or rain to fertilize the eggs of female cones. Just like flowering plants, either one or two plants may be involved in reproduction.

1. *chloroplasts*
2. *vascular plants*
3. *xylem tubes*
4. *root hairs*
5. *root cap*
6. *phloem tubes*
7. *minerals*
8. *petiole*
9. *veins*
10. *stomates*
11. *chlorophyll*
12. *photosynthesis*
13. *light energy*
14. *chemical energy*
15. *nitrogen*
16. *molecules*
17. *cell respiration*
18. *guard cells*
19. *sexual reproduction*
20. *asexual reproduction*
21. *sperm cell*
22. *egg cell*
23. *zygote cell*
24. *gametes*
25. *stamens*
26. *pistils*
27. *pollination*
28. *fertilization*
29. *embryo*
30. *germination*

REVIEW QUESTIONS FOR CHAPTER 7

1. What is transported in xylem and phloem tubes?

2. What are the two important jobs of roots?

3. List three important jobs of stems.

4. What are the two features of leaves that help identify plants?

5. Describe the process of photosynthesis.

6. List three products of photosynthesis.

7. What are stomates, and what do they do for plants?

8. What two vital substances do plants provide for animals?

9. What is cell respiration?

10. What purpose do flowers serve for plants?

11. What is the difference between pollination and fertilization?

12. What is a seed, and how is it formed?

CHAPTER GOALS:

To describe the structure and function of the systems of the human body.

To recognize that the systems work together in the body to carry out the basic life activities.

KEY IDEAS:

Ingestion and digestion make food available and ready for use by body cells.

The circulatory system, the respiratory system, and the excretory system transport food molecules, oxygen, and carbon dioxide and nitrogen wastes.

The nervous system, sense organs, and endocrine glands send and receive information to control and coordinate the body's electro-chemical reactions.

The skeletal and muscular systems provide support, shape, and protection. They work together to produce body movements.

CHAPTER 8

Understanding Systems of the Human Body

Introduction

Imagine the best automobile in the world. It has a frame that is strong and yet lightweight. Its body is sturdy, so strong that it resists most dents and scratches, but smooth and beautiful to look at. Imagine that if it does get a scratch, it repairs itself and looks as good as new. Its gas mileage is very efficient. Its body parts are always well-lubricated. They last until the car is very old. They do not require much maintenance, only basic good care and servicing.

Imagine that carbon monoxide exits through the exhaust system without polluting the air. The automobile responds to external signals and other vehicles. When its oil is running low, for example, it will flash a light, so that the oil can be replaced or changed.

The amazing car that you have just imagined is not nearly as amazing as your body. Like the imaginary automobile, your body has a frame for support and to protect the vital parts inside. This frame is made of bones, which are strong but light. When you get a cut, you bleed, but then the blood stops flowing, and eventually the skin heals itself, usually without leaving a scar.

Your brain and nerves, which comprise your nervous system, take in signals from the outside and respond by doing things such as moving body parts. Better than any automobile or machine ever made, your body is the most marvelous "machine" in the world.

KEY IDEA #1:

Ingestion and digestion make food available and ready for use by body cells.

Ingestion and Digestion

Getting food into your digestive cavity is called **ingestion**. Ingestion is eating and drinking. Breaking down food into liquid form which is simple enough to go into the cells is called **digestion**.

When digestion is over, pieces of original solid food are changed to watery chemicals, so small that they cannot be seen. Even if you looked through the most powerful microscope in the school, you would not see them. If you did see something, you would not even recognize it as food. It would have been changed into chemicals called food molecules.

Digestion Begins Inside Your Mouth

You chew and crush food with your teeth and jaws, turning it over with your tongue. This is called **mechanical digestion** because the size of the food particles is changed. When more than just the size of the food pieces is changed, the process is called **chemical digestion**.

Chemical digestion begins in the mouth. In your lower jaw, under your tongue, and in your cheeks are the **salivary glands**. These glands make a liquid chemical called saliva. Saliva is

Tongue

Salivary glands

an enzyme that changes some of the food. Along with water it changes the food into a simpler form while you chew and swallow.

From the Mouth to the Esophagus

Knowing what happens to the food you eat while it is in your mouth may not seem exciting. It is, however, very important to the body you live in. You probably enjoy eating. Food often tastes good, and sometimes feels good — like cold ice cream, a juicy apple, or crispy crackers. Taste buds in the tongue are sensitive to four tastes — sweet, sour, salty, and bitter. Chewing food and moving it around in the mouth prepares it for swallowing and movement through the **esophagus**, a long tube that connects the mouth to the stomach. Rhythmic movement caused by muscles and nerves begins here. This automatically moves your food along the digestive tract.

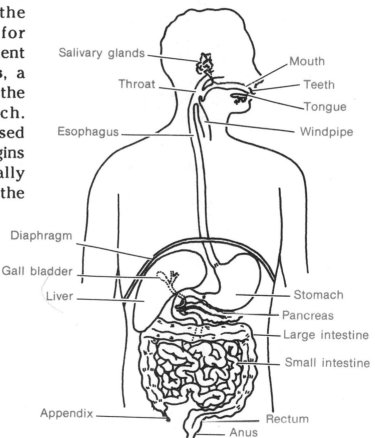

Human Digestive System

The Stomach

Your **stomach** is an organ that stores the bulk of your meal. Digestion takes place as the strong muscle walls contract to mix and churn the food. The stomach wall produces strong digestive juices and a fairly strong acid which acts upon the food while it is being mixed. Solid food becomes more fluid in the stomach. This fluid food is called chyme. Digestive **enzymes** are proteins that help in breaking down food. These enzymes act chemically to separate food materials into smaller molecules. Special enzymes are found in each part of the digestive system. The digestion of protein begins in the stomach.

The Small Intestine

After it leaves the stomach, the food mass goes to the small intestine, the main organ of the digestive system. Over 23 feet long (seven meters), the small intestine does a big job. The main part of digestion is completed in the small intestine.

Inside the small intestine are **glands** that help to digest food. Glands are structures that produce and **secrete**, or give off, special chemicals. Each of the juices or chemicals does a special job. It acts on only one kind of substance. The food that you eat is made mainly of fats, carbohydrates, and proteins. Each part of the food needs to be acted upon and changed by just one kind of enzyme or secretion. The **pancreas** and the gall bladder are glands that get into the digestive act.

The **liver** is a very large organ that secretes **bile**. Bile is stored in the gall bladder, a gland which lies very close to the liver. Bile breaks apart fat molecules. The liver stores useful carbohydrates and fats for short periods. It is a very active organ. In addition to helping with digestion, the liver is very active in energy changes. The liver acts to control the level of a number of substances. The liver changes substances to useful and less harmful forms. The liver breaks apart protein molecules.

All of the juices from the pancreas, gall bladder, small intestine, and the liver complete the breakdown of food. Finally, the food molecules are small enough and changed enough to move out of the digestive tract through the walls of the small intestines. Capillaries, lymph vessels, and small blood vessels pick up the digested food molecules. Food materials that have not been digested move on to the large intestine.

Inside the Large Intestine

Digestion does not take place in the large intestine. The large intestine stores food that is not digested. Water is removed from the food mass and solid wastes are formed. The solid wastes are called feces. Feces are stored temporarily in the last part of the large intestine, the rectum. Feces leave the body through a body opening called the anus.

CHECK YOUR UNDERSTANDING
1. What is the first step in the process of digestion?
2. Tell the main difference between mechanical and chemical digestion.
3. Describe the main digestive organs of the body.
4. What glands and large organs help with the chemical digestion of food while the food mass is in the small intestine?
5. Tell what happens in the large intestine.

KEY IDEA #2:
The circulatory system, the respiratory system, and the excretory system transport food molecules, oxygen, and carbon dioxide and nitrogen wastes.

Body Systems and Life Activities

The respiratory system absorbs oxygen from the blood stream through the lungs. By way of circulation the oxygen goes to body cells. Oxygen in the cells acts

---

on the food molecules. Energy is released from the food. Carbon dioxide gas and water are formed as waste products. The blood carries the waste back to the lungs and out of the body. Carbon dioxide is released through the lungs.

The circulatory system serves as the transport system for food molecules, oxygen, and carbon dioxide. All of the body systems work together. The circulatory system also carries the secretions of the body's glands to the special places where they are needed. The end result of all of the systems working together is that basic life activities are carried out.

Circulation and the Heart

The **circulatory system** in the human body consists of a pump that forces blood all around the body through a network of tubes transporting blood. The pump is the heart, and the tubes are the blood vessels.

The Heart

The **heart** is the most powerful organ in the body. It is made mostly of muscle tissue. The heart contracts and relaxes in a regular rhythm. The rhythmic beat is known as the **heartbeat**. The beat of the heart is *automatic*. The beat is under the control of a separate system of nerves that regulate the rate of the beat. The heart beats about 70 times a minute when at rest. It beats much faster — up to 180 times per minute — when you run, jump, swim, or do aerobic exercises.

Pulse is a regular bulge in the artery walls caused by increased pressure inside the walls as the heart beats. The force of blood in the arteries can be felt as a pulse at the wrist. By holding two fingers on your wristbone below your thumb, you can count the number of times your heart beats in a minute.

How the Heart Works

The heart is about the size of a human fist. It is located between the lungs in the chest cavity. The heart has a left side and a right side. Each side has an upper chamber and a lower chamber. There are four chambers in the human heart.

The right atrium and then the right ventricle receive oxygen-poor blood from the body. The right ventricle is muscular and sends the blood to the lungs.

Parts of the Human Heart

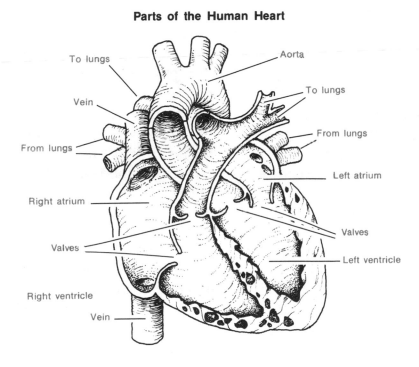

To lungs

Aorta

Vein

To lungs

From lungs

From lungs

Left atrium

Right atrium

Valves

Valves

Left ventricle

Right ventricle

Vein

In the lungs, the blood gets filled with oxygen. From the lungs, oxygen-rich blood goes to the left atrium and then into the left ventricle. The left ventricle is a thick muscle powerful enough to send the blood surging out through the largest of arteries, the aorta, to the entire body. Heart muscle is called cardiac muscle. Its special design transmits pulses on a regular basis.

Blood Plasma

The fluid part of the blood is the plasma. Proteins in the plasma called antibodies fight diseases. When a foreign substance enters your body, your body reacts by making antibodies. Antibodies attach themselves to the invaders by chemical means. This defense system makes an antibody for each kind of invading disease. Antibodies are responsible for body immunity, which is the ability to resist disease.

Red Blood Cells and Hemoglobin

Red blood cells

Half of the blood is made up of red blood cells. Red blood cells are the carriers of oxygen. They contain hemoglobin. **Hemoglobin** is a protein that joins together easily with oxygen. Oxygen plus hemoglobin gives blood its red color. It is an unusual substance because it attracts and carries oxygen in the bloodstream. Hemoglobin also gives up oxygen easily for the cells to use. A normal, healthy body has a red blood cell count of 4.5 to 5.5 million cells per cubic millimeter.

White Blood Cells to the Rescue

White blood cells

White blood cells make up a much smaller part of the blood. There are normally between 7,000 and 10,000 white blood cells per cubic millimeter. Each white blood cell is much larger than a red blood cell. The white cells are the body's protection against infection. Some wrap themselves around bacteria. Others make chemicals which kill bacteria. The white cells are able to move out from the bloodstream to where there is an infection. The white blood cell count increases when a bacteria or virus enters the body.

Blood Vessels and Circulation

Blood vessels are hollow tubes that transport blood to and from all parts of the body. The large vessels are similar to major highways where blood is speeding along at a fast rate. These vessels are **arteries** and **veins**.

The arteries take blood full of oxygen and food to the smallest blood vessels. Arteries become smaller and smaller as they are found further from the heart. Very small arteries branch out into blood vessels called **capillaries**. The capillaries are so narrow that they seem like small paths in the woods compared to the wide highways that are our arteries. The walls of the capillaries are one cell layer thin, and the exchange of oxygen, food molecules, carbon dioxide, and waste materials takes place through the walls. Your body has so many capillaries that each of the millions of cells are next to a capillary wall. Waste products from cell activity are pushed into the capillaries and then into the smallest of veins on their return to the heart and the lungs. Large veins rush the blood along as it returns to the heart for more oxygen.

Blood Vessels

Blood Pressure

Blood pressure is a measurement of how hard your blood pushes against the inside of artery walls and veins. When your heart beats, blood is pushed into the arteries. This forces the artery walls to bulge for a moment. As the large arteries become smaller and smaller, and as blood goes farther away from the heart, more force is needed to keep blood flowing. This makes the heart work very hard. If blood pressure goes up too high, the heart can be damaged.

Platelets and Clotting

Blood platelets are the colorless substances in plasma that are helpful in clotting blood. Their action helps form a network of fibers in the blood plasma when there is a cut. Red cells get caught up in these threads and form a clot. Platelets are unusual because they normally do not get involved in clotting in the bloodstream — only when the rough surfaces of a cut appear. Normally there are about 300,000 platelets per cubic millimeter of blood.

Blood platelets

Blood Types

There are four blood types in the world: Type A, Type B, Type AB, and Type O. The different types are caused by different proteins in the red blood cells. Each person has one of the blood types. Sometimes people who are injured or who have certain illnesses may need more blood. Transfusions are given to replace lost blood. Blood typing must be done to identify the patient's blood type. If the wrong blood type is given during a blood transfusion, clotting takes place. Clotting, or clumping, of red blood cells blocks the tiny capillaries. The normal oxygen delivery system breaks down. Without oxygen, cells die. Blood banks with a variety of blood types are set up to provide needed blood.

The Respiratory System at Work

The work of the **respiratory system** is to attract oxygen from outside your body into your lungs. The lungs are full of small blood vessels that are able to attract oxygen. When you breathe in air, the oxygen moves into your lungs and into your blood vessels.

Lungs are the most important organs in the respiratory system. You have two lungs, one in the right side of your chest, and one in the left side. The lungs serve your body by connecting to the outside air and the inside walls of blood vessels.

When at rest, you usually breathe about 14 times a minute. With each breath, you take in about a half a liter of air. A strong sheet of muscle underneath your lungs separates the lung cavity from the abdominal cavity. It is called the **diaphragm**. The diaphragm is a muscle that helps you breathe. By contracting or pulling, it helps make air move in. It relaxes and moves up, and you breathe out. Your lungs are made of tissue which can stretch. When you **inhale**, or breathe in, the lungs fill up with air. When you breathe out, or exhale, the air in the lungs moves out, and the lungs become smaller. Inhaling and exhaling air mixed with carbon dioxide is a mechanical process. This process depends on changing pressure inside the body.

Air Inside the Lungs

Air comes in through two body openings; your nose and your mouth. The air travels into one large breathing tube called the **trachea**. The trachea branches into two tubes called **bronchi**. One bronchus goes into each lung. Here each bronchus branches out into many tubes called **bronchial tubes**. Inside each lung, these breathing tubes continue to branch and become **bronchioles**.

Finally, at the end of the tiniest bronchioles, there are many sacs of air which are so small that you would have to look through a microscope to see them. They are thin and full of tiny blood vessels. These microscopic air bags inside your lungs are called **alveoli**.

Oxygen: From Respiration to Circulation

Walls of the tiny alveoli air sacs are usually only one cell layer thick. The walls are always moist. Wrapped around the alveoli, just like tiny nets, are many tiny capillaries. Oxygen that is in the air inside the alveoli moves

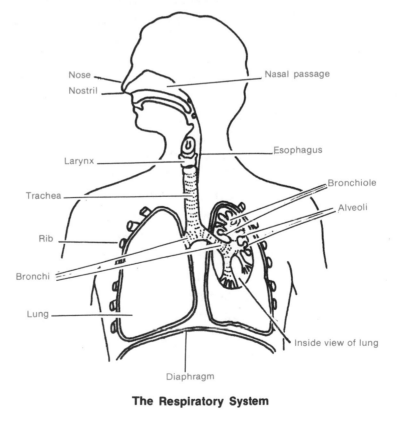

Nose
Nostril
Larynx
Trachea
Rib
Bronchi
Lung
Nasal passage
Esophagus
Bronchiole
Alveoli
Inside view of lung
Diaphragm

The Respiratory System

through the walls of the tiny air sacs, through the walls of the tiny capillaries, and into the blood. Oxygen moves from air inside the alveoli to the blood in the capillaries through a process called diffusion.

The Lungs as Excretory Organ

The lungs are excretory organs. When the blood inside capillaries picks up oxygen in the lungs, it also releases carbon dioxide into the air inside the lungs. Carbon dioxide is what is left after cells in your body break up food molecules. Oxygen acts on the food to release energy for basic life activities. Blood full of carbon dioxide returns to the capillaries in the lungs. Then the carbon dioxide is on its way out of the body.

Kidney Structure and Function

The kidneys serve as excretory organs which remove wastes from the blood. There are two kidneys, and they are located in the small of your back, behind your stomach and below the diaphragm muscle. Kidneys are the most active and important excretory organs. They help maintain chemical balance in the body fluids. They serve as filters to get rid of extra sugar and other waste products in a substance called urine.

Urine Formation in the Kidneys

Plasma from the blood inside the capillaries goes into tiny capillary beds called nephrons. Wastes and extra water from the plasma stay inside the nephrons. Much water, sugar, and minerals from the plasma go back into the capillaries. The wastes which stay inside the nephrons make up urine. Urine is mostly water, with some nitrogen wastes, salts, sugars, and vitamins. Blood continues to go through the capillaries and into veins, which carry it back to the heart. Urine from the nephrons goes into larger collecting tubes in the kidneys to be excreted.

Urine Excretion

The **urinary bladder** is a muscular bag where urine is stored until it is removed from your body. There are two tubes which carry urine from each kidney to the urinary bladder. The bladder stretches as it fills, and then it squeezes together to push urine into a tube called the urethra. The **urethra** is the tube which takes the urine to the outside of your body.

Excretory System (Urinary)

Perspiration Is Formed by Sweat Glands

Your skin is a large organ of your body. One of the important things the skin does is act as a separate system to get rid of extra water, along with some inorganic salts and urea. This waste material is called **perspiration**. There are thousands of tiny sweat glands in your skin. **Sweat glands** are coiled tubes which open up at the outside of your skin. These openings are called **pores**. Perspiration from the sweat glands leaves your body through the pores.

Body Systems Working Together

You have looked at three large systems of the body. It is convenient to study them one at a time, but they work closely together. Circulation, respiration, and excretion are going on at the same time. These three systems all deal with the same substances — blood, food, and carbon dioxide. They are all working together to carry out basic life activities.

CHECK YOUR UNDERSTANDING

1. What are the two main parts of the circulatory system?
2. What do red blood cells and white blood cells do to help the body?
3. What is the main pair of organs of the respiratory system?
4. What is the process by which oxygen gets into the blood in the capillaries?
5. Describe the process of excretion.

KEY IDEA #3:

The nervous system, sense organs, and endocrine glands send and receive information to control and coordinate the body's electro-chemical reaction.

Coordinating Systems of the Body

The nervous system, sense organs, and endocrine gland system act to coordinate and regulate reactions and activities of the body. These systems send messages from one place to another. They tell body parts what to do. They inform the body about what is happening in the outside world. They also regulate body functions because of changes they cause inside the body. They work together to make sure that the body is in balance.

The Nervous System

The human nervous system is efficient and requires little maintenance, although it runs night and day. This amazing human computer consists of the central nervous system and the peripheral nervous system. The central nervous system has two parts—the brain and the spinal cord. The peripheral nervous system consists of all the nerves that transmit messages between the central nervous system and other parts of the body.

Three Parts of the Brain

The three parts of the brain are the **cerebrum**, the **cerebellum**, and the **brain stem**. The largest part of the brain is the cerebrum. The cerebrum controls the way in which you think, learn, remember, and feel about things. Body movement is also controlled by the cerebrum.

The cerebrum is divided into two halves, or hemispheres. The left hemisphere of the cerebrum controls activities on the right side of the body. The right half of the cerebrum controls activities on the left side of the body. The left side of the brain has more to do with math, language, and logical thinking. The right side of the brain has more to do with music, art and other creative activities.

The cerebellum is the smaller part of the brain which is behind the cerebrum. The cerebellum controls balance and helps coordinate muscular activity such as walking and writing. The cerebellum controls these activities by sending and receiving messages between the sense organs and the muscles.

Under the cerebellum is the brain stem, which connects the brain and the spinal cord. A part of the brain stem called the medulla controls the automatic activities of your body such as heart rate, gland secretions, digestion, respiration, and circulation. You do not think with your medulla. It is always on "automatic." Otherwise you would spend all of your time operating the parts of your body.

The Spinal Cord

The spinal cord is a thick bunch of nerves starting at the medulla and going down the back. It is protected inside a spine of bony vertebrae. The brain sends out and receives information through the spinal cord. Thirty-one pairs of spinal nerves branch off from the spinal cord. The spinal nerves send nerve impulses all over your body. The spinal cord, brain stem, and the brain are the central controls of the sense organs and body systems.

Sending and Receiving Messages

Your body constantly receives signals from the outside world through your eyes, ears, and other sense organs. The central nervous system responds to these signals by sending out impulses. Your body responds to these impulses in different ways. It reacts in order to protect itself, to maintain itself, and to continue to do all basic life activities. Some reactions are automatic. Reactions that require thought take more time.

What Neurons Do

Nerve cells are called **neurons**. Information travels from one neuron to the next by an **impulse**. An impulse is a very small electric or chemical charge which moves very fast. There is a short gap between neurons called a **synapse**. Within each neuron an impulse travels from end to end. When an impulse comes to the end of one neuron, there is a chemical change at the synapse: the next neuron picks up the impulse. Information moves through your body by traveling along lengthy sets of neurons.

Three Kinds of Neurons

There are three kinds of neurons in your nervous system; sensory neurons, association neurons, and motor neurons. Sensory neurons send impulses from sense organs to the spinal cord or the brain. Association neurons in the brain and spinal cord carry the impulses from sensory neurons to motor neurons. Motor neurons carry impulses from the brain and spinal cord to muscles and glands.

Reflex Actions

Sneezing, coughing, and blinking your eyes are reflex actions. They happen automatically. What happens if you accidentally touch a frying pan which is very hot? Sensory neurons send the "It's hot!" message from the heat and the feeling it causes. The neurons send impulses to the nerve cord or spinal cord. Inside of the spinal cord, association neurons read and interpret the message and send impulses out to the motor neurons. This is called a reflex arc. All of this happens in an instant. You pull your hand away very quickly. You have been saved from a serious burn by the reflex actions of your autonomic nervous system and muscle system.

In a similar way, if some object comes flying toward your eyes, you blink without thinking! Reflex actions are automatic, and they are fast. They are mostly built-in protective actions.

Sense Organs

The body connects with the outside world through special organs called sense organs. The five main sense organs are the: 1) eyes, 2) ears, 3) skin, 4) nose, and 5) tongue and mouth.

Eyes and Light

The eyes of animals are their cameras to the outside. Highly specialized, they focus light in all kinds of conditions. Light enters the eye through the **cornea**, which is a clear, transparent covering layer that is surrounded by the **sclera**, or "the white of the eye."

The color of your eyes is in the **iris**. The Greek goddess of the rainbow was named Iris. The muscles in the iris contracts the **pupil**, or opening of the eye. This controls the amount of light that enters. Your pupils get larger when you enter a dark room. The larger opening lets in more light and lets you see things more clearly. Likewise when you go into the bright sun, the pupils shut down to very small openings to keep too much light from entering. Too much light or glare sends too many signals to the nervous system, and that blots out the image. The bright light of direct sunlight can damage the receiving screen in the back of the eyeball. You should never look directly into the sun.

Eyebrow

Pupil

Iris

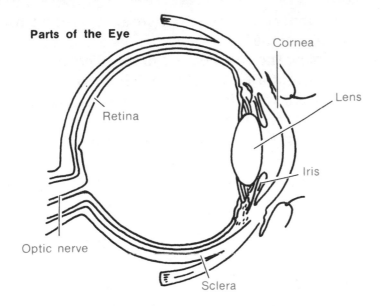

Parts of the Eye

Cornea

Lens

Retina

Iris

Optic nerve

Sclera

The Lens and Retina

There is a clear, soft lens behind the pupil opening in the iris. Muscles on the lens cause it to change its shape. By changing its shape, the lens changes the focus of light rays on the inside wall of the eye.

The retina is the inside wall of the eye and is covered by special receptor cells and nerve cells. The retina receives information from the light rays which focus upon it. The retina sends this information to sensory neurons. The sensory neurons form the optic nerve, which goes to the brain. The brain gives meaning to the impulses from the optic nerve and forms a picture. All of this happens very quickly. Light travels fast, and so do nerve impulses.

The Sense of Hearing

Your **ears** send information about sounds to your brain in the form of nerve impulses. Sounds are collected in the outer ear and are sent into the ear canal to the eardrum. The eardrum is a thin tissue which separates the outer ear from the middle part of the ear. Three small bones in the middle part of the ear send on the vibrations from the eardrum.

In the inner part of the ear there are spaces filled with liquid. One of the liquid-filled spaces of the inner ear is the cochlea, the real hearing organ. The cochlea has hearing receptor cells inside which are like tiny hairs. Sounds from the middle part of the ear cause liquid inside the cochlea to move around. Movement of this liquid bends the hair-like receptor cells. The receptor cells send impulses to the auditory nerve which goes to the brain. The brain receives impulses from the auditory nerve and gives meaning to the sound impulses.

Parts of the Ear

Skin as a Sense Organ

The largest organ of the body is the skin. This organ has receptors that send messages for heat, cold, heavy touch, light touch, pressures, and pains. From the tip of your toes to the top of your head, your skin is able to pick up messages and send them to the nervous system. There are billions of receptor cells in the layers of skin that cover the body. The skin is the body's first line of defense against disease. It is also the first line of defense in warning of any kind of danger, such as hot frying pans, sharp edges, or harsh chemicals.

Taste and Smell

The sense of taste depends on the sense of smell. They both have receptor cells which react to chemicals in food. Taste buds on the tongue pick up four basic kinds of taste: sweet, sour, bitter, and salty.

The sense of smell is less well described. Receptors for smell react to thousands of different chemicals.

The Endocrine Glands

A system of glands known as the endocrine system releases chemical messengers, called hormones, directly into the blood stream. Hormones help regulate what happens in tissues and organs all over the body. Growth and development of body parts is controlled by hormones. Concentrations of different chemicals are regulated by hormones.

The hormone **insulin** from the pancreas regulates the level of sugar in the blood. Without insulin, sugar cannot be absorbed into the cells where it is needed and the sugar level in the blood gets too high. Diabetes, a destructive disease, results.

Too much insulin "burns up" the sugar in the blood, and cells cannot obtain the energy they need to operate. The brain is one of the first organs affected.

In general, hormones speed body reaction rather than slow it down. Keeping an internal balance is an important role that is regulated by hormones from the endocrine system. The body does not store hormones. More than twenty different hormones have an effect on everything from the development of sex characteristics, to kidney function and regulation of stress reactions, such as heart rate and blood pressure.

Basic life activities are regulated in many ways by secretions from the endocrine glands. It is very important to have the correct amounts of hormones at the right time in order for your body to carry on basic life activities.

CHECK YOUR UNDERSTANDING

1. Name the three functions of your nervous system.
2. What are the three parts of your brain?
3. How does information travel in your nervous system?
4. What are the five sense organs of your body?
5. What are the four basic kinds of tastes?

KEY IDEA #4:

The skeletal and muscular systems provide support, shape, and protection. They work together to produce body movements.

The Human Skeleton

The 206 bones of the human body are strong and light enough for ease of movement. They form the skeletal system. The bones give support to the body and provide it with a shape. The mixture of big bones, small bones, flat bones, wide bones, and specially shaped bones of the spinal cord give the body ways to do many different things.

The bones form a framework that surrounds the inner organs and gives them protection. The vital lungs and heart are encased in a rib cage. This protects them from injury.

The bones are circular and hollow. This structure is one of nature's strongest forms. Inside of the bones there is a substance called marrow. It serves as the source of red and white blood cells. Chemicals essential to certain activities, such as phosphates and calcium, are stored and released by the bones as needed.

THE HUMAN SKELETON

Cartilage to Bone

Most bones start out as cartilage. Cartilage is a tough tissue which is not as hard as bone. As your body grows, more and more cartilage is changed into bone. This process goes on all your life, but most changes happen before you are twenty-five years old. The process of replacing cartilage cells with bone cells is called ossification. Not all cartilage changes to bone. Feel the end of your nose. It is still cartilage and will never become a hard, compact bone.

Joints

Joints are the point at which bones meet. Cartilage covers bones at the joints. This cartilage acts like a cushion and protects bones from rubbing against each other. At the movable joints the bones are connected to one another by special tissues called **ligaments.** Ligaments stretch in order to allow the bones to move. At the same time, ligaments hold the bones together.

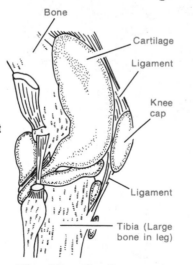

The Knee Joint

Bone

Cartilage

Ligament

Knee cap

Ligament

Tibia (Large bone in leg)

Two important kinds of joints for movement in the skeletal system are the hinge joint and the ball and socket joint. The joint in your elbow is a hinge joint. The joint in your shoulder is a ball and socket joint. Some joints can only move a little, like your rib and spine joints. Some joints, such as those in your knees, move a great deal.

The Muscle System

Muscles work by contracting or tightening and then relaxing or resting. Muscles are for movement. Movement of the arms, legs, and torso are the big movements. Other muscle movements take place in the heart, the digestive system, and in the rib cage and diaphragm.

Most bones have striated skeletal muscles attached to them. These muscles act with the bones. The bones serve as levers to exert forces started in the muscles. The skeletal and muscular systems work in unison.

Muscles are either **voluntary** or **involuntary**. Skeletal muscles, like those in the arms and legs, are voluntary, because you can make them work or rest. You can wiggle your toes or wave your hands. You cannot control involuntary muscle, such as the muscles of your stomach. Involuntary muscles work automatically.

The biceps muscle contracted

Skeletal Muscles Are Voluntary

Skeletal muscles are voluntary muscles which are attached to bones or to one another. You work these muscles to move your bones and make your body active. Skeletal muscles bend and extend. Skeletal muscles cause movement only by pulling on bones; never do they push bones.

The large muscles of the upper arm of humans is a good example of muscles attached to bones and the pulling action that makes the arms move. The biceps muscle is the one in the upper arm that bulges to the front. When showing off muscles, this is the one most often flexed.

Involuntary Muscle

Cardiac muscle is involuntary muscle that is only in the heart. It contracts and relaxes automatically.

Another kind of muscle which contracts and relaxes automatically is smooth muscle. Smooth muscle is involuntary muscle that is found in the walls of digestive organs and blood vessels.

CHECK YOUR UNDERSTANDING

1. What are four functions of the skeletal system?
2. Name a hinge joint.
3. Name a ball and socket joint.
4. How do muscles make things move?
5. What are the three kinds of muscle in your body?

Summary of Chapter 8

The digestive system makes food ready to be used by the cells. Food that is ingested must be made small enough and simple enough to be used by the cells in your body. Digestion makes food into usable food molecules.

The heart pumps blood through vessels to cells in your body. Blood carries food molecules, oxygen, special kinds of blood cells, and chemicals. Oxygen enters your lungs with the air that you breathe. In the lungs, oxygen diffuses into the blood. Carbon dioxide from the cells is breathed out. Your blood, lungs, kidneys, and skin remove wastes from the cells to the outside of your body.

Sense organs send information to the central nervous system. Nerve impulses travel from one neuron to another as tiny electrical charges. Your central nervous system gives meaning to this electric information and sends out impulses on nerves that cause you to react.

Your endocrine glands make chemical juices which they put into your blood. These chemicals regulate and balance the way that parts of your body work.

The muscles and bones in your body support your weight. They give shape to your body while protecting softer organs inside. Muscles relax and contract to make your bones work together like levers, so you can walk, jog, run, or do gymnastics.

Each of the systems of the body has a different function to perform. Yet all body systems work together in order to carry out basic life activities.

1. *ingestion*
2. *digestion*
3. *mechanical digestion*
4. *chemical digestion*
5. *salivary glands*
6. *esophagus*
7. *gland*
8. *secrete*
9. *pancreas*
10. *circulatory system*
11. *heartbeat*
12. *pulse*
13. *arteries*
14. *veins*
15. *capillaries*
16. *hemoglobin*
17. *respiratory system*
18. *diaphragm*
19. *diffusion*
20. *excretory system*
21. *urethra*
22. *urinary bladder*
23. *cerebrum*
24. *cerebellum*
25. *brain stem*
26. *spinal nerves*
27. *neurons*
28. *impulse*
29. *synapse*
30. *cornea*
31. *sclera*
32. *iris*
33. *pupil*
34. *cochlea*
35. *hormones*
36. *insulin*
37. *ligaments*
38. *muscle system*
39. *voluntary muscle*
40. *involuntary muscle*

1. What is the difference between ingestion and digestion?

2. What is the difference between mechanical and chemical digestion?

3. What do the salivary glands do?

4. What happens to food in the stomach?

5. What makes the liver a special organ?

6. What is the main organ in the circulatory system?

7. What is hemoglobin? What does it do?

8. How are capillaries different from arteries?

9. What is a pulse?

10. What part does the diaphragm play in the respiratory system?

11. What do the kidneys do?

12. How does the skin serve as an excretory organ?

13. What is the job of the nervous system, sense organs, and the endocrine system?

14. What is the main difference between a sensory neuron and a motor neuron?

15. What is the job of the sense organs?

16. How does the pupil regulate the amount of light that enters the eye?

17. What is the function of the receptors in the skin?

18. Describe the function of hormones.

19. What does the skeletal system do for the body?

20. What is the difference between voluntary and involuntary muscles?

CHAPTER GOALS:

To recognize how organisms produce offspring like themselves.

To recognize that there are patterns in the processes of reproduction, growth, and development of new organisms.

KEY IDEAS:

The offspring of each animal species are like the parents, but they are not identical.

Sexual reproduction is a duplicating process that transmits genetic material to the next generation.

Organisms go through stages of growth and follow patterns in their development.

Reproduction, growth, and development in humans follow a pattern similar to that of other mammals.

CHAPTER 9

Describing Patterns In Reproduction, Growth, and Development

Introduction

Every species must reproduce in order to maintain a population. Individuals eventually die; but, because of reproduction, the population lives on. Parent organisms create offspring like themselves; but, in sexual reproduction, the offspring are not exactly like either parent. The offspring receive half of their genes from each parent. This method of reproduction creates individuals that are truly unique. Humans and other mammals reproduce sexually.

KEY IDEA #1:
The offspring of each animal species are like the parents, but they are not identical.

Life From Other Life

What would happen to a species that did not reproduce? Sooner or later, every member would die, and the species would be gone—**extinct**. All living species keep themselves going by means of reproduction. In fact, the ability to reproduce is one of the basic life activities for a species.

Spontaneous Generation

It may seem obvious that rabbits produce more rabbits and giraffes produce more giraffes; but, many years ago, some people believed in **spontaneous generation.** They saw flies emerge from rotten meat and frogs hop out of the mud of a pond in the spring, so they reasoned that rotten meat produces flies, and mud makes frogs. Early scientists discovered that eggs laid in meat were the source of the flies and that eggs found on the edges of ponds were the source of the frogs. Spontaneous generation does not exist. Life comes from other life.

DNA Stores the Patterns of Life

Each animal's cells contain the information needed to make another animal just like itself. Most cells contain

chromosomes located in the cell **nucleus**, or center. These chromosomes are made of *DNA*, a most important life chemical. Without DNA, there would be no life at all.

DNA stores information in a pattern, much as a book stores information in a pattern of letters and words. The pattern in a cell can be read by other chemicals to make a new cell. The patterns in the sex cells of rabbits contain all the information needed to make rabbits. Cottontail rabbits and jack rabbits are similar, but their different DNA makes them into either cottontails or jack rabbits.

Most Reproduction Requires Two Parents

One rabbit cannot reproduce by itself. Some very simple animals can, but all the more complicated species require two parents — a male and a female. Because a rabbit has two parents and gets exactly half of its DNA from each, that rabbit is never an exact duplicate of one parent. It may resemble one more closely than the other, but it carry traits from both parents.

The Advantage of Diversity

Variety or **diversity** gives the population better chances to adapt. It is much better for the survival of a species if the population is made up of individuals with a range of differences. In case of a disaster, there is a much greater chance that some individuals in that species will have traits that allow them to meet with changes in the environment. These individuals can survive to carry on the species. Individuals in the population that can not adapt to sudden environmental changes will not be able to survive.

Adaptation and Resistance

Suppose a disease sweeps through a population of deer, killing many members. If none of them is able to fight the disease, then all would die. Chances are that a few, at least, will be resistant to the disease and will survive. You can say that resistance is a kind of adaptation. Adaptations come from different combinations of genes and DNA.

Insects have vast numbers of offspring, each slightly different from the next. Often they become a problem in farm areas by eating crop plants. Spraying chemicals never seems to kill all of the pest insects. No matter what spray is used, some seem to be resistant to its effects. Of course, those are the only ones who live to reproduce. The next season, those insects with resistance to the spray reproduce offspring like themselves. More of the population is resistant to the chemicals. After a while the chemical spray that was once effective becomes completely ineffective. A new chemical must be developed because the insects became resistant to the first chemical.

CHECK YOUR UNDERSTANDING

1. What is the purpose of reproduction?
2. What is spontaneous generation?
3. What is DNA, and what does it do?
4. What advantage is there to having variations among individuals in a population?
5. What causes you to resemble your parents and your grandparents more than you resemble other people?

KEY IDEA #2:

Sexual reproduction is a duplicating process that transmits genetic material to the next generation.

Sexual Reproduction

Sexual reproduction is the formation of a new individual from the union of two cells. It is the type of reproduction that most animals use. Sexual reproduction involves the duplication of the chromosomes of the cell and the division of the cell itself. Although cells copy themselves and divide all the time, in sexual reproduction the process is not the same.

Cells Divide to Make New Cells

The basic structure of living things is the cell. A common feature of a cell is that it divides. Skin cells, bone cells, muscle cells, and the lining cells of body organs must all reproduce themselves. The control center in the process is the nucleus. In the nucleus of each cell are sets of chromosomes. There is a set number of chromosomes for each species. Fruit flies have four pairs of chromosomes, while humans have 23 pairs.

Chromosome Duplication

The first step in the making of a new cell is the copying or duplicating of the old cell's DNA in its chromosomes. Each new cell will need a complete copy. Division into two new cells can proceed only when gene materials are in the right amounts. For a short time, just before cell division occurs, a cell has two complete sets of chromosomes.

Mitosis

After the chromosomes have been duplicated, they line up at the center of the cell. Soon each double pair of

Steps in Mitosis

chromosomes separates, one pair going to each end of the cell. Then the cell membrane pinches in between two sets of chromosomes, separating the original cell into two identical or "daughter" cells. Each daughter cell has an exact copy of the chromosomes present in the parent cell. The process of mitosis is now complete.

Asexual Reproduction

If an organism consists of only a single cell, mitosis reproduces the entire organism. This process is called **asexual reproduction**, meaning reproduction without sex, because only one parent is involved.

Asexual reproduction has one advantage — speed. A bacterium or other single-cell protist can reproduce every twenty minutes or half-hour. The chief disadvantage of asexual reproduction is its lack of diversity. All the offspring are exact copies that will respond to changes in the environment in the same way. If there is a dangerous chemical in the environment that kills one of the organisms, it will probably kill them all.

Sexual Reproduction in the Animal Kingdom

The majority of animal species in the world use **sexual reproduction**. Sexual reproduction involves the genetic materials from two individuals. This guarantees more diversity and variety of traits and characteristics than in asexually produced organisms. All of the vertebrates and almost all insects use sexual reproduction. These groups are the most successful organisms in the world.

Gametes from Gonads

All higher animals reproduce sexually using two parents of different sexes, male and female. The female produces **egg cells** from an organ called an **ovary**. The male produces **sperm cells** from organs called the **testes**. The sex organs, the ovary and testes, are called **gonads**.

Nucleus

Egg cell

Sperm cell

The sex cells that the gonads produce, the egg and the sperm, are called **gametes**.

The testes of male animals produce millions of sperm cells. Egg cells are rather large as cells go, but sperm cells are very small, about one thousandth the size of an egg cell. Eggs contain food for the early stage of the developing offspring. Sperm are mostly a nucleus, with a half set of chromosomes and mitochondria to produce their energy for movement.

1.

2.

3.

Nucleus

Chromosomes lining up

4.

5.

6.

Chromosomes separating

7.

8.

9.

Two cells

Dividing again

Four daughter cells

Steps in Meiosis

Meiosis

Both egg cells and sperm cells are single cells containing only half as many chromosomes as the parents. These gametes, or sex cells, are formed by a process called meiosis (my-oh-sis). Meiosis is similar to mitosis, but with more dividing. Producing eggs and sperm is more complicated.

As in mitosis, the first step in meiosis is the copying and duplicating of the DNA. The cell divides in the usual way. Then the cell divides a second time *without* duplicating the DNA. Four daughter cells are formed, each with half the usual amount of chromosomes. This set of daughter cells becomes the egg cells and in the male, sperm cells. Meiosis is used only to produce gametes, so it occurs only in the ovaries and testes.

Fertilization

Like the egg, each sperm has half the usual number of chromosomes and cannot survive on its own. Unlike the rather large, round, and slow-moving or stationary egg, the sperm propels itself forward by means of a tiny tail which it rapidly whips back and forth. Many species produce millions of sperms at one time. Although all the sperms move toward an egg cell, only one sperm can fertilize the egg. One sperm gets through the cell wall of the egg to join with the egg, fertilize it, and create a new cell. The nucleus of the egg cell and the nucleus of the sperm cell each provide half the chromosomes. Together, the egg and the sperm provide the newly formed cell with a full number of chromosomes. A full number of chromosomes is essential to making a new individual.

The Zygote Is a Fertilized Egg

Imagine an egg of a fish attached to a rock on the bottom of a stream. A female fish has laid her eggs. Each egg is a tiny speck, but it carries the possibility of a new fish within it. It is surrounded by hundreds of other eggs just like it. Above the mass of eggs swims a male fish. He releases sperm cells into the water, billions of them, and they begin a frantic race against time. Each swims through the water toward an egg. One sperm reaches an egg just ahead of dozens of competing sperm. The successful sperm drives itself right through the outer

membrane of the egg, piercing this layer. (All animal cells have a containing structure called a membrane.) The membrane quickly seals the puncture. Almost immediately, the membrane changes so that no other sperm can come through. Unused sperm and unfertilized eggs do not survive.

Once inside the egg, the sperm continues to move and meets the chromosomes of the egg cell. The chromosomes from the nucleus of the sperm and the egg unite. The egg is now fertilized. The fertilized egg, called a **zygote**, begins to develop into a new fish.

External and Internal Fertilization

The process of fertilization described above is called **external fertilization** because the egg is fertilized outside the body of the female. Most fish and amphibians use external fertilization, but some fish and all reptiles, birds, and mammals use **internal fertilization**, in which the male puts sperm inside the female's body, where fertilization occurs.

CHECK YOUR UNDERSTANDING

1. What is the first step in mitosis?
2. What is a daughter cell?
3. What is the main disadvantage of asexual reproduction?
4. What is the difference between mitosis and meiosis?
5. What is a zygote?

Organisms go through stages of growth and follow patterns in their development.

Growth and Development of Animals

An organism begins as a zygote and goes through stages of development until it becomes an adult. Cells duplicate and divide to form new cells. The organism grows. If it is a mammal, when it reaches a certain developmental stage, it is ready to live outside of the mother's body.

From Zygote to Embryo

A zygote is a single cell that contains a full set of chromosomes. Many things must happen before the zygote becomes an adult. In the case of a fish, the zygote divides to form two identical but smaller cells attached to each other. This process is repeated many times, forming four, then eight, then sixteen cells. In a few days, there are millions of cells. The zygote grows into an embryo. An embryo is an early stage in the development of an organism.

| Fertilized egg | Two cells | Four cells | Eight cells | Many cells |

Development

Gradually, the new cells that are formed take on different shapes and functions. This is called **development**. Cells at one end of the developing embryo begin to form part of the head — eyes, mouth, gills, and other organs. Those on one side become nerve tissue that eventually make up the spinal cord. Inside, the heart and other organs develop. Signals in the chemical DNA cause these differences in the cells. This is **cell differentiation**. Eventually, a complete fish is formed. It is tiny (no bigger than the original egg), but cell differentiation has resulted in all the parts necessary for a fish. Not just any fish, but a fish of the same species as the parents.

Growth

This young fish, still curled inside its egg, has a difficult job ahead. It is so small that its chance of survival is not good. Other organisms in the environment compete for food, and developing eggs are a good target. The fish must grow rapidly or die.

The fish hatches and begins to eat. Anything small enough to go in its mouth is fair game — algae, baby shrimp, any kind of microscopic water creature it can catch. It, too, is food for larger fish and other animals. Many of its species become meals for turtles, birds, and other fish. As it eats and grows, its chances for survival improve. If it escapes being food for another organism, it will live long enough to be able to be a part of the reproduction of the species. Most of the hatch are not so lucky. That is why fish must lay thousands of eggs in order for a few to survive. Successful reproduction, development, and growth are essential for a species to survive.

Development in Mammals

Mammals are different from fish in many ways, but their development begins the same way. A big difference is that mammals take better care of their offspring for longer periods of time. Protection from danger and preparation for living by themselves give mammal offspring a better chance to survive. Therefore, fewer individuals need to be born to ensure survival of the species.

Like all other animals, mammals produce eggs and sperms. The vast majority of mammals do not lay eggs. The egg cells develop and are fertilized inside the body of the mother. There are only two egg-laying mammals; the spiny anteater, and the duck-billed platypus. Both animals live in Australia.

Stages of Development

Some offspring look like their parents right from the start. A newborn calf, a young colt, or a baby giraffe are not mistaken for some other animal. The tadpoles that develop from frog's eggs do not look like frogs at all.

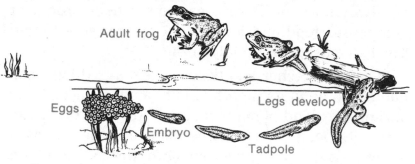

Butterflies, moths, and beetles go through stages where they are not at all like adult insects. Butterflies and moths have a caterpillar stage and a cocoon stage. Beetles develop from the pupa stage into adult beetles.

Egg　　Larva (caterpillar)　　Pupa (cocoon)　　Adult butterfly

Grasshoppers and praying mantises hatch as small grasshoppers and mantises. They are called nymphs.They just get bigger and bigger, but look like their parents right from the start. Babies of human beings look like miniature humans for the most part. They just grow bigger, become children, go through some major changes during their teens, and reach maturity as young adults.

Larger Offspring

Mammal babies may seem small and cuddly to us, but they are gigantic compared to a baby fish, which is smaller than a fingernail. The fish must depend on the small food supply in its egg, but the mammal embryo gets food from its mother throughout its development. For this reason, the mammal can grow much larger and develop more fully before its natural birth.

The Mammal Uterus

Every mammal female, from cats and dogs to horses, whales, and people, has a **uterus** inside her body. The uterus is an organ designed to hold, feed, and protect a developing embryo, and to expel the fetus when its development is complete. The egg, fertilized by the sperm from a male, develops in the uterus. As the fetus develops, it gets food from the mother and is also protected because it is inside her body.

Inside the uterus, the fetus forms protective membranes around itself, making a fluid-filled sac for the embryo. Parts of these membranes form a **placenta**. A placenta is a special organ that gets food from the mother's uterus and passes it to the developing embryo. Most mammal species form placentas.

Food for the Young Mammal

A mammal egg is much smaller than a hen's egg or even a fish egg, because it does not need a food supply for more than a few days. The food that it gets from the mother is much more than could ever be packed into an egg.

After birth, the mammal offspring continues to get nourishment from the mother in the form of milk. No other type of animal cares for and feeds its young as mammals do. As a result, most mammals have few offspring because they can expect many of them to survive.

Gestation Differences

All mammal embryos follow the same basic pattern of development. However, the **gestation** time required varies greatly depending on the size of the animal. Gestation time is the time from the fertilization of the egg until birth. Gestation times for a range of mammals is shown in the following chart.

MAMMAL GESTATION TIMES

Mammal	Approximate Number of days
Mouse	20
Rabbit	31
Cat or Dog	63
Monkey	210
Human	275
Cattle	281
Horse	336
Whale	360
Elephant	624

A whale baby takes almost a year to develop, and an elephant more than a year and eight months. That is a long time for a mother to carry a baby! In general, the larger the mammal, the longer the time required for development.

Some animals are exceptions. For instance, the opossum, which is larger than a cat, has a gestation period of only 13 days. That is because an opossum is a **marsupial** mammal. In marsupial animals, the embryo in the uterus gets no food from the mother. Instead it must be "born" at a very early stage. Then, the embryo goes into a pouch on the mother's body and nurses milk for a long time while it completes its development.

CHECK YOUR UNDERSTANDING

1. What is cell differentiation?
2. Do young frogs and young grasshoppers resemble their parents?
3. Why do fish need to lay thousands of eggs at one time?
4. What does the uterus do for the developing embryo?
5. What is meant by gestation time?

Human Reproduction, Growth, and Development

Humans are mammals with the same reproductive organs as other mammals. Humans use sexual reproduction and internal fertilization. Like other mammals, the human embryo develops inside the uterus of the female. After birth, the baby gets nourishment from the **mammary glands**, or breasts, of the female. In fact, mammals get their name from these mammary glands which produce milk after the birth of a child.

The Male Reproductive System

The **testes** and the **penis** are organs of the male reproductive system. The testes produce the male sex cells called sperm cells. The testes are contained in a sac called the **scrotum**, which lies outside of the main body wall. Sperm are sensitive to heat, and this location, outside the body proper, provides some relief from the higher temperature inside the body. Like humans, monkeys, horses, and cattle have testes which are outside the body. Other animals such as frogs, birds, and snakes have testes which are inside of the body cavity.

Sperm cells travel through a tube in the penis called the **urethra**. The **prostate gland** also connects to the urethra. The prostate gland provides additional fluid called **semen** which mixes with the sperm cells and carries them through the urethra. The semen then flows through the urethra, which travels through the penis to the outside of the body. Urine also flows through the urethra of the penis to the outside. However, urine and sperm do not flow through the urethra at the same time.

Male Reproductive System

Labels: Urinary bladder, Sperm duct, Penis, Urethra, Scrotum, Testes, Rectum, Seminal vesicle, Prostate gland

The Female Reproductive System

Eggs are produced in **ovaries,** as in all other animals. The two ovaries are located inside the abdomen, below the stomach. One egg cell is produced at a time, usually about every 28 days. The exact time varies with the individual, but **ovulation,** or egg production, occurs about once each month.

From the ovary where it is formed, the egg cell travels through a tube into the uterus, or womb. If sperm are present, the egg cell is fertilized in the tube. If it is not fertilized, the egg dies and passes out of the body.

Female Reproductive System

Fallopian tube — Ovary

Uterus

Urinary bladder

Bone of pelvis

Rectum —

Urethra

Vagina

Menstruation

Each month, the uterus prepares to receive a zygote. Its walls thicken to form a soft, blood-rich cushion waiting to hold and nourish an embryo. If the egg cell is not fertilized, the lining of the uterus breaks down and is discharged from the uterus through the vagina. Most of the discharge is blood that was in the tissues of the lining. This process is called **menstruation** and is necessary so that a fresh lining can be prepared for the next egg cell a month later.

Basic Biology

Human Fertilization

Like all mammals, humans use internal fertilization. The penis, the male reproductive organ, is used as the means of getting sperm to the egg in the body of the female. In many species, the penis is concealed in a small cavity, but in humans it is external.

In preparation for fertilization, blood flows into special vessels in the penis, extending its length and making it rigid. Only when erect can it be inserted into the female's vagina, an opening which leads to the uterus. Semen, which contains sperm, together with fluid from the prostate gland is released into the vagina. Sperm cells must then swim to the uterus under their own power. If an egg cell is present, one of the sperm cells breaks through the egg membrane to fertilize it. The female is now pregnant.

Pregnancy

If the egg cell is fertilized, the wall of the uterus catches it and gives it a home. The embryo forms a

Placenta

Umbilical cord

Fetus

Uterus

Vagina
(Birth canal)

Pregnancy at Nine Months

placenta, which meshes with the *uterine wall* like a hand with thousands of fingers inside a glove made to fit it perfectly. The mother's blood flows through a capillary bed in the uterine lining, and the baby's blood flows in a capillary bed through the placenta.

The two blood supplies do not mix, but they come so close together that food and oxygen pass from the mother's blood to the baby's. Waste products are exchanged for food and oxygen, and the mother gets rid of the waste products. Her blood does not flow through the baby directly.

The placenta is connected to the embryo by a cord called the **umbilical cord**, which contains blood vessels. The embryo is well developed by the age of five weeks. The umbilical cord is tied off at birth and becomes the "belly button," or navel.

The Embryo Becomes a Baby

A fetus at 6 months

Inside the uterus, the embryo develops rapidly, using the food supply provided by the body of the mother. The embryo first forms a hollow ball of cells. Quickly, the cells become different in appearance and function. The body takes shape, and organs appear. **Differentiation** and **development** continue to change cells to tissues and tissues to organs.

After about three weeks, the heart begins to beat. Blood vessels form rapidly, and the body clearly shows a head, buds for arms and legs—and a tail. The embryo is still smaller than a fingernail.

Basic Biology

The human embryo at three weeks looks similar to the embryo of a fish, chick, or horse. (All animals with backbones look very similar in their early stages of development.) A human embryo soon moves on from this primitive stage.

At about four weeks, tiny hands begin to show fingers, and eyes start to show up as dark spots. By the time pregnancy is half over, at 4-1/2 months, the fetus shows thumb sucking behaviors inside the womb. (The embryo is called a fetus in the later stages of development.) Sucking is an instinctive behavior which the newborn baby can use to suck milk from the mammary glands.

Birth of a Baby

When the fetus reaches full size, most often after nine months of development, the uterus begins to contract. Over and over, the uterus squeezes to push out the baby. Slowly at first, every half hour or so, then gradually faster until contractions come about every five minutes, the process of delivery goes on.

After much work, the mother has a new baby in her arms. The umbilical cord is still attached at the navel, and the placenta is pushed out soon after the baby is born. The cord is clamped and cut, and the baby can survive outside the mother's body.

Parental Care

The parents care for and feed their new offspring for many years. Compared to a baby horse, which stands in just a few hours, runs the next day, and stops nursing after only a few months, the human baby is still helpless. There is no chance that human offspring could take care of themselves in less than several years.

Much time and effort must be devoted to caring for babies and young children. The basis of the human family is the caring and nurturing of children. Other animals have hundreds of thousands of offspring at a time and then leave them without care. Humans are different. Human beings usually have one child at a time and often several years apart. The care that they give their children is one reason for the high survival rate of humans.

Adolescence

Humans care for their offspring through the special period of the teens called **adolescence** and often to adulthood. Adolescence is a period of increased development in humans. No other species has such an extensive and lengthy period of growing up. Growing up brings with it some special and rapid physical changes. These changes take place during **puberty**, which occurs at the beginning of adolescence.

Changes in Males and Females

In the case of males, there is a change in the voice from a high to a low pitch. Hair begins to grow on the face, under the arms, and in the **pubic** region where the sex organs are located. The sexual organs become more fully developed, and the production of sperm is highly increased.

Females also undergo changes during puberty. Pubic and underarm hair start to grow. The mammary glands enlarge, and the menstrual cycle begins. Adolescents are fully capable of reproduction.

Not only is adolescence a period of physical growth, it is also a time when there is increased emphasis upon individual behavior and the assumption of responsibility. In humans, the period of adolescence is the developmental stage in which the child becomes more and more like an adult—and very often more like the parents.

1. What is the purpose of mammary glands?
2. Describe the path followed by sperm from the testes to the site of fertilization.
3. What is menstruation?
4. What does a three-week-old human embryo look like?
5. What are some changes that occur in males during puberty?

Summary Of Chapter 9

A species becomes extinct if it does not reproduce. Reproduction ensures that traits of the parents are passed on to the next generation. The traits of an individual lie in the chromosomes of the cells. Chromosomes are long strands of DNA which give the cell and the individual its hereditary qualities and characteristics.

Cell Duplication

Cells generally reproduce by mitosis. In this process the cell first duplicates its DNA. Then it divides in half. The daughter cells are identical to the parent cell.

In sexual reproduction, two parents are involved. In a process called meiosis half of the genetic material comes from each of the parents. This material goes to the egg cell and to the sperm cell. An advantage of sexual reproduction is that all of the genetic material does not come from just one parent. It creates great diversity among offspring. The offspring are not exactly like either parent.

Reproduction, Growth, and Development

Although there are many different animals in the world, all species follow similar patterns of growth and development. When a sperm cell goes through the membrane of an egg cell, a zygote is formed. A zygote is a fertilized egg. It begins as one cell, then it divides and differentiates until it is becomes an embryo with body parts and organs.

In some species, the female lays eggs that are fertilized outside her body. In mammals, the egg is fertilized internally and develops inside the body of the mother. Since the mammal nurtures and protects the developing embryo, more individual mammals survive.

Human Reproduction

Like other mammals, humans fertilize their egg cells internally. Developing embryos are protected in the uterus and are nourished from the mother's body. After birth, babies are fed milk and are cared for by their parents for many years. In humans, there is a period of development called adolescence, in which children gradually become adults.

1. extinct
2. chromosome
3. spontaneous generation
4. adapt
5. diversity
6. gene
7. DNA
8. resistant
9. sexual reproduction
10. egg cell
11. gonad
12. fertilize
13. external fertilization
14. embryo
15. nucleus
16. asexual reproduction
17. sperm cell
18. gamete
19. zygote
20. mitosis
21. ovary
22. testes
23. meiosis
24. membrane
25. reproduce
26. internal fertilization
27. gestation
28. uterus
29. cell differentiation
30. placenta
31. marsupial
32. scrotum
33. penis
34. semen
35. development
36. urethra
37. ovulation
38. vagina
39. adolescence
40. mammary glands
41. prostate gland
42. womb
43. umbilical cord
44. puberty
45. menstruation
46. fetus
47. pubic

REVIEW QUESTIONS FOR CHAPTER 9

1. Why must living organisms reproduce?
2. Why do brothers look different from each other?
3. What is DNA and why is it important?
4. How does adaptation give a species a greater chance for survival?
5. How is DNA copied?
6. Why do insecticides only work well for a few years?
7. What are chromosomes?
8. Describe the process of mitosis.
9. What are gonads?

10. What are gametes?
11. What is the difference between mitosis and meiosis?
12. What is an example of asexual reproduction in animals?
13. What is the advantage of sexual reproduction over asexual reproduction?
14. Where does meiosis occur?
15. What is the difference in chromosome number in gametes and other body cells?
16. What is the difference between a zygote and an embryo?
17. What is cell differentiation?
18. Describe how external fertilization occurs in fish.
19. What is the difference between the way frog babies and human babies develop?
20. Why must fish lay thousands of eggs?
21. What is gestation time?
22. What is a placenta?
23. How is fish reproduction different from human reproduction?
24. Why can only one sperm fertilize an egg?
25. Name two ways that humans are like other mammals.
26. Why do humans produce relatively few babies?
27. What evidence is there in the development of an animal embryo that many animals have a common background?
28. How does a baby get out of the uterus?
29. What is the belly button?
30. Why are human testes outside the body wall?
31. What happens to a human egg that is not fertilized?
32. What is the difference between internal and external fertilization?
33. What is ovulation?
34. What are some changes that occur in females during puberty?

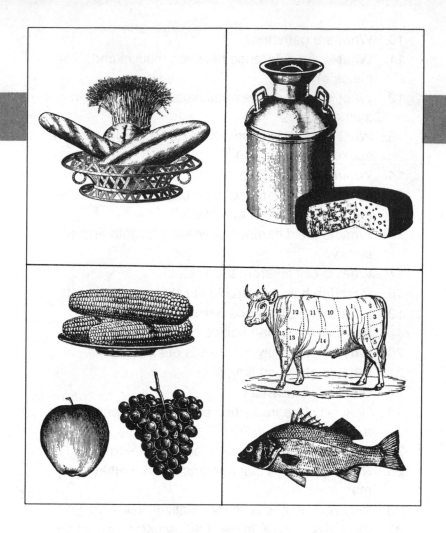

CHAPTER GOAL:

To discover how disease and diet each have a part in the human fight to stay healthy.

KEY IDEAS:

The human fight to stay healthy is the story of progress in the control, prevention, and cure of diseases that affect the body.

Good nutrition and a balanced diet are major factors in the human fight to stay healthy.

Understanding The Human Fight To Stay Healthy

The Human Fight Against Disease

Infectious Diseases

At one time, **infectious** diseases were responsible for wiping out whole populations. In the 1300s, two-thirds of the people of Europe were stricken by the bubonic plague, or Black Death, as it was called. In the year 1665, more than 30,000 people in London died of this disease. In the 18th century, one out of every three English children was infected with smallpox, and those who survived were scarred with pockmarks. In the 19th century, a plague of another dread disease, cholera, wiped out thousands of people in the United States. Even in the twentieth century, polio, a disease of the nervous system, was a killer and crippler of children. In the United States and other industrialized countries, these diseases have been practically eliminated with the process of immunization.

Germs

At one time people thought that the wind or evil spirits were responsible for bringing infectious diseases that killed whole populations. Now it is known that germs or **pathogens**, disease-causing materials, are responsible. These germs are usually so small that they can only be seen with a microscope. For this reason they are called **microbes**. Microbes can be protozoans, bacteria, or viruses. These microorganisms can be carried from place to place by mosquitoes, flies, fleas, rats, and pigs, and then transmitted to humans. Microbes can be found in water, waste products, unwashed silverware, and in the coughs and sneezes of people who are sick. Microbes are everywhere.

The Body's Defense Systems

If microbes are everywhere, why aren't you sick all the time? You stay healthy because you have a series of defense systems that prevent germs from entering your body. The first lines of defense are the skin, the lining of the nose and throat, the hairs of the nostril, and the tonsils. They contain lymph nodes that have many white cells. The acids in the stomach also kill bacteria. If the microbes get past these first lines of defense, they are surrounded by white blood cells. The white blood cells capture the germs before they can do any harm. Even

with those two defense systems, some microbes do get into the body. What happens then? That's when the body's immune system takes over. Lymphocytes and leukocytes are white blood cells that are the body's main defense against viruses, bacteria, and other foreign organisms.

Antibodies

Have you ever heard someone say, "Oh, I can't catch measles. I'm immune to them. My brother had them last year, and I slept in the same room, but I never caught them." That person probably has a natural immunity to measles. You might have heard another person say, "I'll never get mumps, I just had a mumps shot." The second person has an acquired immunity to mumps. He or she has been vaccinated or inoculated against mumps.

Your immune system system is your most important defense against disease. When a foreign substance enters your body, your blood forms **antibodies** which help combat the disease. These antibodies help you recover and prevent future attacks. When you get a "shot," you are inoculated with small amounts of a mild form of the disease. This inoculation makes your body produce the antibodies to fight the disease. When your body has produced antibodies against the disease, you will not be able to catch it. You have been immunized against this disease. You can say that you have an acquired immunity because of the inoculation.

Vaccination

At one time, smallpox was a dreaded disease. In the early 1800s, people were inoculated against smallpox by being injected with a small amount of the disease. Unfortunately, some people died as a result of this inoculation. An English doctor, Edward Jenner, noticed that milkmaids who caught cowpox, a mild illness, never developed smallpox. He got the idea that if he introduced a small amount of cowpox into a person's body, that person would be immune to smallpox.

Jenner put cowpox material taken from a cowpox sore on the scratched arm of a young boy. The boy developed cowpox and recovered. Next, Jenner inoculated the boy with smallpox. The boy did not get smallpox at all. The cowpox material had caused the boy's body to produce antibodies. The boy was immune to smallpox. Jenner called his method **vaccination**. Today, because of Jenner's work and the success of his vaccine, smallpox has been eliminated from most parts of the world.

Vaccinations have helped people to become immune to smallpox, measles, mumps, polio, and other infectious diseases.

Sanitation

If you are thirsty, you can get a drink of water. You don't even think about it. You just turn on the faucet and get a cold drink. Previously this was not the case. People drank water that was piped in from rivers in which filth and sewage were present. Sometimes they drank water from town pumps where the water was **contaminated** by cesspools. Waste products in the cesspools and rivers filled the water with bacteria and viruses. There were epidemics of typhoid fever and cholera. Thousands of people grew ill and died. Better sanitation methods and practices are a leading cause of improved health.

No one at that time knew about microorganisms. They did not realize that germs in water or food could infect masses of people. Even medical doctors did not know that germs could travel from one patient and infect the next patient. Louis Pasteur, a French scientist, was one of the first scientists to study microbes.

The body's immune system uses white blood cells to act against foreign substances in the blood. These white cells secrete antibodies. Antigens are foreign substances that are part of bacteria and viruses. They cause an immune reaction. White blood cells respond.

Today, people know that some microorganisms can cause disease. They are careful to keep things clean and sanitary. For that reason they do not drink water or eat food that has been heavily contaminated by bacteria.

The water you drink is treated before it ever reaches the faucet. It is tested by government agencies to make sure it is fit to drink. Food in the supermarkets and medicines in the drug stores have to be safe before they can be sold. Milk has to be pasteurized, or heated in a special way, to make sure the bacteria are destroyed.

CHECK YOUR UNDERSTANDING

1. Name some microbes that cause infectious diseases. Tell how these diseases are transmitted.
2. What are the first two lines of defense of the body against disease?
3. What are antibodies? How do they fight disease?
4. What is done to make sure that our water is fit to drink?
5. What are antigens?

Nutrition and Diet

People often say, "Good nutrition is important if you want to stay healthy." Nutrition is all about how the body uses food. Foods are **nutrients** because they nourish the body or feed it. Nutrients are required for body growth and maintenance. They are assimilated and used by the cells.

Your cells function properly when they have the substances that they need. These are: carbohydrates, fats and oils, proteins, water, **vitamins**, and **minerals**. Each of these substances is used in a special way by the body. Carbohydrates are made of sugars and starches and are the energy producers. That's why you have so much energy after you eat a candy bar, which is full of sugar. Fats and oils give you energy. Proteins are tissue builders. This is the material that the cells mostly use to repair themselves and to reproduce.

Water makes up about 70% of your body tissue. It is the material in which all the body substances are dissolved. You lose water when you sweat. Lungs expel water vapor. You also lose water when you get rid of body wastes during urination. Your body needs a great deal of water. You need to ingest enough liquids so that you won't dehydrate, or dry out.

Vitamins

Vitamins are substances that are found in tiny amounts in animal and plant foods. The vitamin chart on the next page shows how essential vitamins are.

Vitamin and Source	What It Does	Deficiency/Disease
Vitamin A Milk, egg yolk, beef liver	Helps skin, hair, eyes, and lining of nose and throat; prevents night blindness	Dry, rough skin, lowered resistance to respiratory infections, poor vision in twilight.
Niacin Liver, meat, whole wheat, milk	Protects skin and nerves, aids digestion.	Mental depression, digestive disturbances
Vitamin B (Thiamine) Milk, pork, egg yolk, cereal, vegetables, fruit	Protects health of nervous system, aids appetite and digestion.	Tiredness, loss of appetite.
Vitamin B$_2$ (Riboflavin) Milk, pork, liver, eggs, vegetables, fruit	Increases body resistance to infection and prevents harmful changes to eyes.	Eye fatigue, lower vitality.
Vitamin C (Ascorbic Acid) Tomatoes, most citrus fruits	Helps form bones and teeth; prevents scurvy.	Scurvy, sore joints, tender gums, poorly formed bones and teeth.
Vitamin D Fish liver oils, milk, sunlight	Prevents rickets, uses calcium and phosphorus to build bones and teeth.	Rickets, poorly formed bones and teeth
Vitamin E Vegetable oils, cereals, eggs	Probable role in reproduction.	Not known.
Vitamin K Green leafy vegetables, soybeans, bran	Aids in clotting of blood.	Bleeding.

Minerals

Minerals are essential to good health. They are found in very small amounts in the foods that you eat. Calcium and phosphorus build strong bones and teeth. Sodium and potassium help the cells function. Most people get enough minerals in the foods that they eat. Good food must be eaten each day, since minerals are not stored in the body.

Some people, especially older people, have too much sodium. Sodium is table salt, and, as people grow older, they are often told to cut down on salt. They might need a "salt-free" diet, especially if they have high blood pressure. Salt causes your body to retain water in the blood vessels and the cells. Although exact causes of high blood pressure are unknown, many doctors regar salt as a major factor.

The Four Basic Food Groups

Doctors suggest that you eat some foods from each of the four basic food groups. Your body will get the necessary nutrients if you select food from each of the four food groups on a daily basis.

FOUR BASIC FOOD GROUPS

Food Group	Foods In Group	Daily Requirement
Milk Group	Milk, butter, cheese, ice cream	4 glasses or their equivalent
Meat Group	Meats, poultry, eggs, peas, dry beans, nuts	2 or more servings
Vegetable-Fruit Group	Dark green and yellow vegetables, citrus fruits, tomatoes	4 or more servings
Bread-Cereal Group	Bread, cereals, crackers, spaghetti	4 or more servings

Summary of Chapter 10

Infectious diseases are caused by pathogens. Most pathogens are so small that they can only be seen under a microscope. The pathogens are called microorganisms or microbes. Microbes can be bacteria, viruses, or protozoans. At one time, microbes were responsible for widespread death and disease.

Doctors began to realize that microbes were the cause of disease. For this reason they advised people to drink clean water and to live under sanitary conditions. In time, it was discovered that people would develop an immunity to a disease if they were given a mild form of the disease itself. Inoculation caused the production of antibodies to fight the disease. Mass inoculation enabled people to become immune to infectious diseases.

Nutrition is the process of ingesting, digesting, absorbing, and assimilating food. Nutrients in food are necessary for cell repair and growth. In order to function, the body needs carbohydrates, fats, proteins, water, vitamins, and minerals. A balanced diet supplies these substances. Eating food from the four basic food groups (meat, milk, grain, and fruit and vegetables) gives the body the nourishment it needs.

1. *vaccination*
2. *antibody*
3. *contaminated*
4. *infectious*
5. *microbe*
6. *pathogen*
7. *antigen*
8. *dehydrate*
9. *vitamin*
10. *mineral*
11. *nutrition*
12. *nutrient*

REVIEW QUESTIONS FOR CHAPTER 10

1. What are pathogens?

2. What is a microorganism? What instrument is used to view microorganisms?

3. How do white blood cells prevent you from becoming sick?

4. How did Edward Jenner fight disease?

5. How do inoculation and sanitation prevent infectious diseases?

6. What is nutrition?

7. List four of the six substances that your cells need in order to function properly. Tell why each substance is necessary.

8. How can salt affect blood pressure?

9. Why is it important for you to eat proteins?

10. Name the four basic food groups. List two foods from each of the four groups.

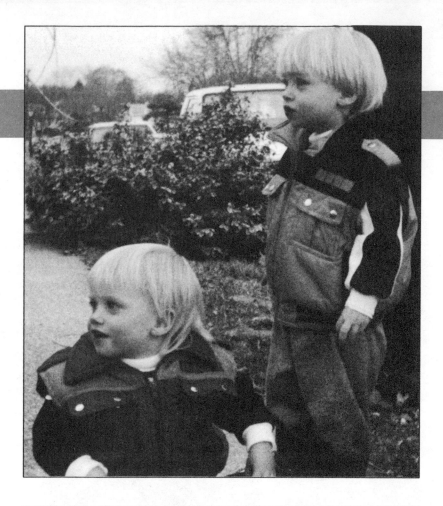

CHAPTER GOAL:

To examine and understand how traits and features of one generation of organisms are passed to the next generation.

KEY IDEAS:

The transmission of traits from one generation to the next occurs in a regular pattern.

The basis of heredity is what chromosomes do during sex cell formation.

Human genetics is the study of heredity as it applies to humans.

Applied genetics makes it possible to change the traits of organisms.

11

Following Genetics From Generation to Generation

Introduction

Heredity

The way you resemble your parents, from the freckles on your nose to the length of your legs is due to your heredity. The laws of heredity apply to all organisms, whether they are rose bushes or spiders or frogs. Special cells from parents contain the chemicals that determine what a new individual will be like.

The human species, by means of sexual reproduction, produces unique, one-of-a-kind, persons. Think of the great variety of human beings around you. You can see that each person is different from the next person. Your fingerprints are one of a kind. That is why fingerprints are used to identify a person.

Environment

Heredity does not stand alone. There is always an environment which affects the original material. A tomato seed may have inherited the ability to produce the greatest tomatoes in the world. But, without an environment of sunshine, proper soil, water, and favorable temperature, it will not become as good a tomato plant as it might with the right conditions.

Genetics

Genetics is the study of heredity. Genetics is about the basic laws of heredity and how these laws work. By studying genetics, you can find out about human heredity and how you came to be the way you are. Genetics also includes the science of improving plant and animal species.

KEY IDEA #1:

The transmission of traits from one generation to the next occurs in a regular pattern.

What Is Heredity?

Family members have some of the same traits. The passing of traits from parents to their children is called heredity.

Have you ever been told that you look like someone in your family? Have you ever observed that people in a particular family often look alike?

All organisms pass their traits to their children or offspring. Racehorses resemble other racehorses. Plow horses resemble other plow horses. Siamese cats produce Siamese cats. Collie dogs have litters of collie puppies.

Gregor Mendel

Over one hundred years ago, a scientist named **Gregor Mendel** made important discoveries about heredity. Mendel took seeds from tall pea plants and planted them. Most of the plants from the "tall" seeds were tall, but some of the plants were short. Mendel wanted to find out why the short pea plants were unlike their tall parents.

Pure Traits

Mendel began his investigation by growing plants that were pure for the trait of tallness. Plants that were pure for this trait would only produce tall plants from their seeds. He did this by the method of **self-pollination**. Flowering plants have both the male and female sex organs in their flowers.

Mendel transferred the pollen from the **anther**, the male sex organ, to the **pistil**, the female sex organ. In this way, the ovary could be fertilized, and seeds would be produced. Mendel used these seeds to grow pure, tall pea plants.

Using the same method of self-pollination, Mendel grew plants that were pure for the trait of shortness. He called the "pure tall" and the "pure short" plants the P_1 **generation**. P_1 stood for the first or parent generation of pure plants. To understand what a generation is, think of your grandparents as one generation, your parents as another generation, and yourself as still another generation.

The Next Generation

Mendel wanted to find out what a pea plant would be like if it had one pure tall parent and one pure short parent. He used the method of **cross-pollination** to provide two different parents. Mendel transferred the pollen of pure tall pea plants to the pistils of pure short pea plants. He also took the pollen of pure short plants to fertilize the ovaries of pure tall plants.

Parents
(P_1)

Cross-pollination

Offspring
(F_1)

Pure tall   Pure short   Hybrid tall

Mendel's Experiment

Then Mendel took the seeds from the cross-pollinated plants and planted them. He called the new plants the F_1 generation. The "F" stood for filial, or son.

To his surprise, Mendel found that all the new plants in the F_1 generation were tall. Mendel crossed thousands of pure short and pure tall plants. He observed the seeds and the new plants carefully. But, each time, the results were the same. All the plants were tall. What had happened to shortness? Why had this trait disappeared?

Punnett Squares

A special arrangement called a **Punnett square** can be used to show Mendel's findings. A capital T is the symbol for the trait of tallness. A small t is the symbol for the trait of shortness. TT stands for a pure tall pea plant, and tt stands for a pure short pea plant.

When Mendel crossed pure tall pea plants with pure short pea plants, he produced an F_1 generation of all tall plants. You can see by the symbols in the boxes that shortness did not disappear. It is still there. But it does not show up, because shortness is a factor of less influence.

Mendel's factors are called genes. A factor that is often hidden is a **recessive** gene. A factor that does show up is called a **dominant** gene. A recessive gene does not produce an effect when it is paired with a dominant gene. Plants that have more than one kind of gene for the same trait are called hybrid plants. The plants in Mendel's

F_1 generation were hybrids. A gene for tall and a gene for short were paired, Tt.

Genotypes and Phenotypes

A hybrid has two different genes. One gene has a stronger influence, and the other gene has less influence. The pea plants in the F_1 generation were tall, because tallness was caused by the gene partner with the stronger influence. However, inside the pea plants there was a combination of genes for tall and short in pairs. What is inside an organism, in its genes, is called the **genotype**. What the organism looks like — its appearance — is its **phenotype**. All organisms have both a genotype and a phenotype.

The F_1 Generation

Mendel worked in his monastery garden and produced another generation of pea plants by self-pollinating the hybrid or F_1 generation. He called this third group of plants the F_2 generation. He went from P, to F_1 and now F_2, three generations of painstaking gardening and record keeping. This time, Mendel found that short plants had returned. In the F_2 generation, 3/4 of the plants were tall, and 1/4 were short. Mendel repeated his experiments and always got the same results. The percentage in the F_2 generation was always 3/4 tall and 1/4 short.

The following Punnett square shows what the F_2 generation was like. Three fourths of the plants were tall, and one fourth were short.

What percent of the plants were pure? What percent were hybrid?

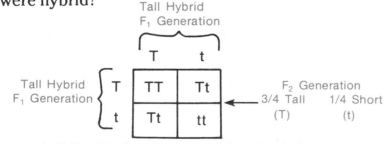

Mendel's Laws

Mendel also checked other traits such as seed color, seed texture, pod appearance, pod color, and position of the flowers. Each trait followed the same kind of pattern as the one for tall and short. Round seeds were dominant over wrinkled seeds. Yellow seeds were dominant over green seeds. Green pods were dominant over yellow ones. Mendel concluded that there must be some factor in a plant that causes it to have certain traits such as tallness or shortness. He reasoned that since two factors are involved, they must come in pairs. Mendel also thought that one factor must be more powerful than the other factor, since the factor of stronger influence sometimes hides the factor of less influence. Mendel called the stronger factor "dominant" and the weaker factor "recessive."

How Color is Determined in Certain Flowers

Mendel also thought that the paired factors must separate during reproduction, so that the new egg receives half the factors from one parent and half from the other parent. Today, these factors are called genes, and the study of heredity is called genetics. In Mendel's experiments, genes were responsible for producing short or tall pea plants. Genes give you your blue eyes or your brown hair. Genes are responsible for carrying traits from one generation to the next.

CHECK YOUR UNDERSTANDING

1. What is heredity?
2. What is a pure trait?
3. What did Mendel do to produce a plant that had one tall parent and one short parent?
4. What were the dominant and recessive factors that Mendel found? Why did he call them dominant and recessive?
5. What is genetics?

KEY IDEA #2:

The basis of heredity is what chromosomes do during sex cell formation.

What Chromosomes Do

Chromosomes are long, rod-shaped bodies located in the nucleus of each cell. Chromosomes are DNA that carry hereditary information. Genes are located on the chromosomes. There are gene pairs for every trait in an organism. Heredity is determined by the way the chromosomes divide during the formation of gametes or sex cells.

Nucleus with chromosomes

Body Cell

Mitosis and Cell Division

From your study of simple organisms, you know that amebas reproduce by splitting in half. The nucleus of the ameba has chromosomes that duplicate. During division of the nucleus, two sets of chromosomes are made. The process is called mitosis. The two new amebas each get a set of chromosomes and are exactly like their parent. The ameba is able to duplicate itself because the chromosomes in its nucleus go through mitosis. The whole cell also divides.

In the process of mitosis, first the chromosomes in the nucleus duplicate themselves. Then the nucleus membrane dissolves, and half the chromosome materials go to form one new nucleus. The other half of the chromosomes forms a second nucleus. The cell itself divides, and each cell then has a nucleus with a complete set of chromosomes and genes exactly like the original set. Since the chromosomes are the same, the new amebas are the same. In amebas, parent and offspring are look-alikes.

Not only do one-celled animals carry out the process of mitosis and cell division, higher organisms do, also. Body cells are constantly dividing as you grow. But, in higher organisms, body cells divide by mitosis and cell division only for growth.

Higher organisms reproduce sexually. Two parents are involved. Each parent contributes sets of chromosomes in a male or female gamete. Gametes are egg and sperm sex cells.

Reduction Division

In sexual reproduction, half the chromosomes come from one parent, and half come from the other parent. If sperm sex cells and egg sex cells just followed the pattern of regular cell division, they would have double sets of chromosomes in their nuclei. Then when fertilization of the egg cell occurred, the beginning organism would have two sets or twice as much genetic material as needed by the species. (Remember, each species has a set number of chromosomes.) If carried to the next generation, it would even be more irregular. So that doesn't happen at all.

Doubling of chromosomes for sperm sex cells and egg sex cells is prevented by a special process of reduction during division. Reduction division occurs during meiosis (my-oh-sis), or **sex cell formation.** Reduction division takes place only in the formation of eggs and sperms. The process involves several steps.

Examples of pairs of chromosomes of a human male

Each sex cell divides, so that it contains 1/2 of the species number of chromosomes. When a zygote is formed by the union of the sex cells, their half number (half the chromosomes from each parent) becomes a whole (1/2 + 1/2) number that is the species number.

Human beings have 46 chromosomes — twenty three pairs. If the 46 chromosomes in the sperm cell got together with the 46 chromosomes in the egg cell, a beginning individual would have 92 chromosomes, twice as many chromosomes as the normal species number. Because of reduction division, the sperm cell contains 23 chromosomes and the egg cell contains 23 chromosomes — one-half the species number. When a new embryo is formed, it then has 46 chromosomes, 23 from each parent. Forty-six is the normal number of chromosomes for all humans.

The mix of chromosomes from two parents during sexual reproduction produces a completely different organism. The combination of genes that is made from the genes of two parents is a major feature of sexual reproduction. Organisms resulting from sexual reproduction are one of a kind. Each individual has a unique set of chromosomes.

Mendel's Work Rediscovered

By the beginning of the 20th century, chromosomes could actually be seen under a microscope. Scientists had not paid much attention to Mendel's experiments before this. They realized what his work had proven. They concluded that what Mendel had called factors were the genes that carried traits from one generation to the next. At this time, biologists also reasoned that one chromosome came from the female parent and the other from the male parent, and that there were opposite forms of a gene, called **alleles**, such as tall or short for a pea plant. This was called the chromosome theory.

Sex Chromosomes

Biologists can observe the **sex chromosomes.** Humans have 23 pairs of chromosomes for a total of 46. In 22 of the pairs, there are two chromosomes that look very much alike. There is one pair of chromosomes that is different in males and females. These are called the sex chromosomes. Sex chromosomes determine the sex of an individual. They give male or female characteristics to people.

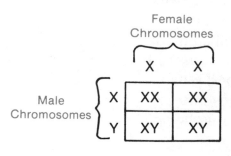

In mammals and flies, the female sex chromosomes match. They are called the X chromosomes, and are represented by the XX symbol. In males, the chromosomes do not match. They are called the XY chromosomes.

The Punnett square shows how the sex chromosomes determine the sex of an individual. The chances are 50/50 of producing either a female (XX) or male (XY).

Experiments with Fruit Flies

Drosophila, a fruit fly with red eyes, has been the subject of many genetics experiments. The *Drosophila* is a good subject for many reasons. Its cells have four pairs of chromosomes, and they are easy to see, since they are huge for such a small fly. *Drosophila* reproduces quickly, is inexpensive to feed, and can be raised in small jars or vials. In addition, it is easy to tell the male from the female. The two-name system holds true for these main players in the study of genetics. They are *Drosophila melanogaster*. No other flies admitted!

Thomas Morgan, a scientist famous for his study of genetics, used *Drosophila*

Drosophila melanogaster

for his experiments. He noticed that one of the male flies had white rather than red eyes. He mated a white-eyed male parent with a red-eyed female parent in order to produce an F_1 generation. All the offspring in the F_1 generation had red eyes. Morgan concluded that red eyes were dominant (Rr) in fruit flies. Mendel had reached the same kind of conclusion about tallness in pea plants (Tt).

Sex-Linked Traits

Next, Morgan mated the flies from the F_1 generation to produce an F_2 generation. Again, the results were the same as Mendel's. Except for one astonishing factor! All the white-eyed flies were males! Morgan concluded that white eye color in *Drosophila* is a sex-linked trait.

Females have a pair of X chromosomes, and are represented as XX. The gene for white eyes is located on the X chromosome from the female. There is no allele for eye color on the male chromosome (XY). The trait for white eyes can only be passed by a female to a male. That is why it is called a sex-linked trait. An organism that has a trait in its genes and passes it on to another generation is called a **carrier**. Being a carrier, the female *Drosophila* never has white eyes herself. However, she can pass down this trait to her male offspring.

CHECK YOUR UNDERSTANDING

1. What are chromosomes?
2. What is mitosis?
3. What is reduction division?
4. What is the difference between the sex chromosomes and other pairs of chromosomes? Write the symbols for the female and male sex chromosomes.
5. Why was the *Drosophila* used for many experiments with genes?

Basic Biology

Human genetics is the study of heredity as it applies to humans.

Human Genetics

Studying human genetics presents more problems than studying heredity in a fruit fly. 1) Human beings have more chromosomes than a fruit fly. 2) Humans do not reproduce as quickly, and 3) Humans cannot be used in laboratory experiments. Scientists study human heredity by observing how heredity and environment affect identical twins. They look at whole families and see how genetic diseases are passed from one generation to the next.

The Gene Pool

To study human genetics, scientists look at samples from the **gene pool.** The gene pool is all the genes in a certain population, such as the population of the United States, or all the genes in a population of white- tailed deer. Scientists can observe specific traits in one generation. These traits are likely to show up in the same proportion in the next generation. A large number of people immigrated to the United States from a great variety of places, so the gene pool of this country is large and varied. Other countries have kept a more uniform population. Their gene pools are less varied than that in the United States.

Studies of Twins

Scientists observe twins to study human genetics. There are two types of twins, identical twins and fraternal twins. Identical twins have duplicate genes, because they come from a single embryo. First, the sperm cell fertilizes the egg cell. Then the zygote divides into two separate cells. Two identical organisms are produced from the separate but similar cells.

Fraternal twins are not identical. They are just like ordinary brothers and sisters. Two different eggs are fertilized by two different sperm cells at the same time. The result is two different offspring with completely different sets of genes.

Heredity and Environment

You are born with certain genes. That is your heredity. Your genes give you your skin color, your eye color, your basic body structure, and other basic attributes. But, you live in an environment that contains all the things around you. Your family, your school, your friends, your total surroundings, all make up your environment. **Geneticists** often study identical twins who have been separated since birth. Both twins have the same genes, but they have grown up in different environments. Geneticists make comparisons and observe the effect of heredity and the influence of environment on a person's development.

The Influence of Environment

Poor nutrition, drugs and alcohol, and smoking can affect an individual. The same factors in the internal environment during pregnancy can affect the development of an embryo. Studies show that babies born to mothers who smoke cigarettes have lower birth weights than babies born to non-smoking mothers. Children of heroin addicts are addicts, themselves, when they are born. They have to be withdrawn from the drug after delivery. X-rays can also damage an embryo and cause changes or **mutations** in the genes.

DNA

Mutations can cause abnormalities in humans and other organisms. The white-eyed fruit fly was a mutation. But how do these genetic changes occur?

DNA is the substance in cells that contains the genetic material that is passed down from one generation to the next. All the information necessary to carry on the life activities is in the DNA molecules. The information is like a code that tells the cells what to do. A major feature of DNA is that it can **replicate**, or duplicate, itself. Each DNA molecule has the ability to split or come apart. It can be compared to a "twisted ladder" that comes unzipped at the rungs (chemicals). New pieces of chemicals join each side of the ladder and an exact copy of the old ladder is made. Now there are two new DNA strands.

This special ability to copy itself makes DNA one of the most important chemicals in living things. Each individual has a separate DNA code. That is why you are different from every other individual.

DNA Molecule

1. DNA Molecule, Showing "Zipper" Arrangment

New chemical materials are selected here.

2. When the zipper separates, each side takes on a new side. The new chemicals attach to the old sides and form new strands.

3. Each of these double strands contains one old strand and one newly replicated strand.

4. The rungs are the genetic code.

Genetic code

Old   New   New   Old

Mutations

Sometimes there is a change in the DNA code. Usually an error occurs in the copying process. The change, called a mutation, can occur from unknown causes, or it can be the result of an environmental factor such as exposure to radiation.

Some mutations are harmful, and others are beneficial. Many of these changes are small and do not have any great effect at all. Some harmful genes are recessive. However, if two persons with the same recessive genes for a particular disease marry, their children can inherit a genetic disease.

Genetic Diseases

Diabetes is an inherited disorder caused by a recessive gene. The body is unable to manufacture **insulin**, which controls the amount of sugar in the bloodstream. People who have diabetes have an abnormal amount of sugar in their bodies. Diabetics can take artificial insulin, either by mouth or by injection, to lower the amount of sugar in their bodies.

Sickle-Cell Anemia is a genetic disorder which affects the red blood cells. The disease results from a pair of recessive genes that are inherited from parents who are carriers. The carriers have the trait for sickle-cell anemia in their genes, but do not have the disease, itself. The sickle-cell trait is carried by eight to ten percent of the black population in the United States. People with this disease have severe anemia which causes weakness and irregular heart action. At present, there is no cure for sickle-cell anemia.

Sickle cells   Normal cells

Abnormalities Through Inbreeding

With the billions of people who live in the world, there are a great variety of genes. A person who is a carrier of a genetic disease is not likely to marry someone with the same genetic makeup. Their children will probably be normal. However, people within a small geographic area often marry each other. Their offspring, again, marry within the area. So, there is a greater likelihood, after several generations, of these couples having children with genetic diseases. The same thing is true when people marry very close relatives. Marrying people within a small gene pool is called **inbreeding**.

Hemophilia

Hemophilia is an example of a genetic disease caused by inbreeding. In hemophilia, the gene that controls blood clotting is defective. Hemophiliacs can bleed indefinitely, and need transfusions of normal blood to help their blood clot.

Hemophilia is sometimes called the "Royal Disease" because many kings and princes in Europe suffered from it. People from royal families generally married other people from royal families, including cousins. Since the gene pool is very small, a recessive trait such as hemophilia shows up readily in the next generation.

Sex-Linked and Sex-Influenced Traits

Did you notice that the last paragraph about hemophilia said, "many of the kings and princes in Europe suffered from it"? The sentence did not mention queens or princesses. Hemophilia, like white eyes in a fruit fly, is a sex-linked trait. It is carried by the female who passes it to the male offspring. Color-blindness is another sex-linked trait. People with this condition cannot see certain colors. The most common type of color blindness is the inability to tell the difference between red and green. Mothers are carriers of color blindness, and pass the trait to their sons.

1. What is a gene pool?
2. What is the difference between heredity and environment?
3. What are some factors that can affect the development of an embryo?
4. What is DNA? What does it do?
5. What is a genetic disease? How is it passed from one generation to the next?

KEY IDEA #4:
Applied genetics makes it possible to change the traits of organisms.

Applied Genetics

Plant and animal breeders use selective breeding to improve organisms. Geneticists transfer desirable genes from one bacteria to another. Scientists are applying their knowledge of genetics to affect heredity.

The Survival of Useful Mutations

Are all mutations bad? Mutation may sound like a bad word, but it has caused many beneficial changes. Mutations in a population can increase the survival of a species of organisms. For example, a mutation that produces an animal with white fur would be helpful in a snowy region. Animals with white fur would probably survive longer than those with dark fur, because they would blend in with the background. Eventually, there would be more white-coated animals in snowy regions. Over long periods of time, mutations cause changes in populations, and new species are formed.

Selective Breeding

Farmers and biologists use their knowledge of mutations to produce new varieties of plants and animals.

Suppose a desirable mutation arises, such as sheep with short legs. Short legs may not seem to be a desirable trait in sheep. However, short legs prevent sheep from jumping over fences. This mutation actually did arise, and short-legged sheep were bred until a new generation of short-legged sheep was produced. The method of selecting useful mutations and breeding organisms so that the mutation shows up again is called selective breeding.

Farmers use selective breeding to produce cows that give great amounts of milk and chickens that lay eggs in larger numbers and larger sizes. Selective breeding has also produced the pink grapefruit and the navel orange.

Breeding Racehorses

Selective breeding is used to produce great racehorses. Good genes are what the horse breeders refer to when they talk about bloodlines in horses. Secretariat won the three most important horseraces in the United States. To breed a winner, Secretariat's owner selected parents with the right bloodlines. Secretariat's father was known for fathering horses that won short races. From his mother's side, Secretariat inherited strength and the ability to go long distances. The result was an oversized horse with a stride of about 25 feet and with excellent speed. Breeders also talk about a horse having a good temperament. Behavior, too, is genetically influenced.

Racehorses are bred selectively.

New Techniques in Genetics

Scientists are able to examine chromosomes under the microscope and photograph them. They arrange them in pairs to see if there are any abnormalities. Sometimes they find a defective gene or an extra chromosome. As recently as the 1970s, scientists began to study genes by introducing new genes into an organism. This is called genetic engineering.

Biologists were first able to transfer genes from one species of bacteria to another species of bacteria. Now they are able to transfer genes between entirely different organisms. This discovery has enabled them to produce substances that are important in medicine, such as insulin, which is necessary for the treatment of diabetes.

So far, scientists have not been able to replace defective genes in humans with normal ones. However, research in this area is still ongoing. Guidelines for controlling genetic engineering experiments have been set by the scientists themselves.

CHECK YOUR UNDERSTANDING

1. What is a desirable mutation? Give an example of a mutation that is helpful.
2. Describe the process of selective breeding.
3. How did Secretariat's genes enable him to be a great racehorse?
4. Define genetic engineering.

Summary of Chapter 11

Heredity is the transmission of traits from one generation to the next. Heredity occurs in an orderly pattern. Many years ago, Gregor Mendel, while working with pea plants, discovered that traits in organisms are due to paired factors, half of which come from each parent. Mendel also found that some factors are dominant and others are recessive.

Mendel's factors are now called genes, and the study of heredity is called genetics. Genes are located on chromosomes which are found in the nucleus of each cell. Chromosomes are able to duplicate themselves, so that each new cell is an exact copy of the original. During the formation of sex cells, the sex cells divide in half. This process is called reduction division. Half of the chromosomes from each parent form the new organism.

Mutations are changes in the genetic material. Harmful mutations cause genetic diseases such as diabetes and hemophilia. Individuals can inherit a genetic disease if they receive the same recessive gene from both parents. Inbreeding causes an unusually large number of abnormalities. With inbreeding, the advantage that a large gene pool provides is lost.

Some mutations are helpful. In selective breeding, the results of mutations are used to produce better varieties of plants and animals. Today, scientists can transfer genes from one species to another. This process is called genetic engineering.

1. *genetics*
2. *Gregor Mendel*
3. *self-pollination*
4. *anther*
5. *pistil*
6. *generation*
7. *cross-pollination*
8. *Punnett square*
9. *recessive*
10. *dominant*
11. *genotype*
12. *phenotype*
13. *chromosomes*
14. *allele*
15. *sex chromosomes*
16. *Drosophila*
17. *carrier*
18. *gene pool*
19. *geneticist*
20. *mutation*
21. *replicate*
22. *diabetes*
23. *insulin*
24. *inbreeding*
25. *hemophilia*

1. Describe the factors that cause you to resemble your parents.

2. What kind of plants were produced when Mendel crossed tall pure pea plants with short pure pea plants? Why did Mendel get this result?

3. What were two of Mendel's four conclusions when he finished his experiments with the pea plants?

4. What is a Punnett square?

5. Write the symbols for:
   a) a tall, pure plant.
   b) a short, pure plant.
   c) a tall, hybrid plant.

6. What happens in the process of mitosis?

7. What happens in the process of meiosis?

8. What traits do sex chromosomes determine?

9. What was Morgan's conclusion when he found that only the male *Drosophila* had white eyes?

10. What are the problems in studying human heredity?

11. What is the difference between identical twins and fraternal twins?

12. Why do scientists use identical twins to study human heredity?

13. What are some environmental factors that can harm the embryo?

14. What is diabetes?

15. Explain why a large gene pool helps prevent genetic diseases.

To find out how organisms interact with the living and non-living parts of their environment.

To discover how natural substances are used and recycled, and what happens to energy in the ecosystem.

KEY IDEAS:

Ecosystems are sets of communities in which organisms interact with each other and with the non-living factors. Ecology is the study of ecosystems.

Food chains and food webs include the food and energy relationships in an ecosystem.

Some substances are reused in a series of cycles as organisms carry out basic life activities.

Energy is essential to all life. Energy constantly enters and leaves the earth and cannot be recycled.

CHAPTER 12

Tracing Cycles In The Environment: Ecology

Introduction

Living things depend on each other and on the natural materials in the world around them. Air, water, and soil, along with energy from the sun, provide the basic needs of land plants and protists. Green plants and protists, in turn, provide for the basic needs of animals. When you study all of the living and non- living parts of ecosystems and how they relate to each other, you are studying **ecology**.

Studying ecology means learning about the environment and its interactions. Ecology includes the study of the cycles in which natural substances are used and reused. Ecology also emphasizes that energy from the sun is essential to all life. The main feature of ecosystems is the continued and high level of interaction between organisms and their living and non-living environments.

Ecosystems are sets of communities in which organisms interact with each other and with the non-living factors. Ecology is the study of ecosystems.

Ecosystems

Organisms do not live alone. They act on and are acted upon by the natural substances in their surroundings or **environment**. Ecosystems are communities of organisms — plants, animals, and protists — interacting with their environment.

The Organism

An organism is a complete living thing. You are an organism. Within you are the cells that form your tissues and organs. All about you are the living and non-living things with which you interact. You interact with the air when you inhale oxygen and exhale carbon dioxide. You interact with plants when you eat the food and use the oxygen they provide. You interact with all the organisms who share the earth and its ecosystems with you.

Populations

Any group of individuals of one species living in one area makes up a **population**. The grizzly bears in Yellowstone National Park are one population. So are the deer in the Shenandoah. The redwood trees in California are another population. All of the people in the United States make up its human population. The bass in the pond form a population of interbreeding fish.

Populations are constantly changing. As people are born and die, the population of this country changes every day. Robins eat worms and raise new robins, changing both the populations of worms and of robins.

Habitats

The place where a population of organisms lives is its habitat. Each population is specially suited to live under special conditions. Bullfrogs live on the edges of lakes, but not in the middle. Tuna fish live only in the salty water of the ocean. Rainbow trout form a population that is specially suited for life in a habitat that is a swift-running part of a fresh-water stream.

Communities

Throughout the world are many populations that live in **communities**. Bears and rabbits and fir trees and grass are populations of organisms, but they can all live together in the same forest. This forest is their community.

A community is a group of populations of different species living in one place. A farm pond is a community of many species. It may contain cattails, algae, turtles, ducks, and several kinds of small fish. Even the worms wriggling in the soil are part of the community.

The populations in a community interact in many ways. The leaves fall from the trees to the forest floor, where some organisms cause decay and return of material to the soil. Squirrels and trees help each other — some of the nuts the squirrels will bury will grow to become new trees. A hawk living in the forest may eat the squirrel and use the tree as a nest. Large trees with thick foliage determine how much light gets through to other plants. Many of the members of the community affect each other.

Ecosystems

An **ecosystem** is the interaction of communities of organisms with their non-living environment. The study of ecosystems is called ecology.

A pond ecosystem is a somewhat closed system. Bass, sunfish, and perch live among the plants, such as pond lilies, duckweed, and arrowhead. Protists like green algae and diatoms store energy of the sun. A turtle captures a fish, and a water snake eats a frog. An ecosystem is a beehive of activity.

Though water flows in and out, carrying nutrients, oxygen, seeds, and some animals, most of the food eaten by the animals in the pond comes from the plants in the pond. All of the living things in the pond ecosystem require energy to live. The energy enters the ecosystem from the sun. Energy stored in plants goes to small and then larger animals.

In the pond, minerals such as calcium are dissolved by the water, absorbed by the plants for their growth,

Basic Biology

and transformed into bones of fish, frogs, and turtles when these animals eat the plants or each other. When an animal dies, its body decays, and the minerals are released into the water to be absorbed by plants all over again. Nothing is ever lost or wasted. Recycling takes place.

Succession

Ecosystems change over periods of time. While ecosystems are not built to last forever, some are less stable than others, and are changed more quickly. Many ecosystems go through a process of change called **succession**, often becoming entirely different.

A pond is an unstable ecosystem. Eventually it may fill up with leaves and soil washed in by a stream that feeds it. After it fills up, it will become a meadow. Fish and cattails will no longer live there, but mice and grasses will. Soon bushes will shade out the grass, then trees will overgrow the bushes. The unstable pond has changed to a meadow. The meadow, too, will change. Larger plants, and then trees, will take over.

An oak, maple, hickory, and beech forest is a climax community. This forest is the last stage in the succession. It may survive for many years. A volcano, a forest fire, or an earthquake might destroy large parts of the forest. The stage is set for some steps of succession to repeat.

Ecology

Ecologists are people who study ecosystems. They may study ponds to find out how all of the living and non-living parts are interrelated and work together. They may want to know which types of fish are able to live and reproduce in the pond.

They may add nutrients or plants or snails to see what effect the change will have on the system as a whole. Ecologists may study other ecosystems such as forests, mountains, or deserts.

Pollution in Ecosystems

People produce a variety of waste and waste products that affect the balance of nature in unexpected ways. Sometimes these wastes pollute water, air, and soil and interfere with natural succession. **Pollution** is any undesirable change in an environment. Pollution is most often connected with human activities.

Lakes are often affected by pollution. Topsoil that is washed off from construction sites and improper farming practices fills up streams and lakes. Fertilizer washes off, polluting the water with chemicals that make undesirable plants in the lake grow faster. Industrial wastes dumped into the water kill fish and cause damage. What happens in one ecosystem affects other ecosystems.

The Biomes of the World

The world has many ecosystems. Some are in the ocean. Others are in the desert. There are different

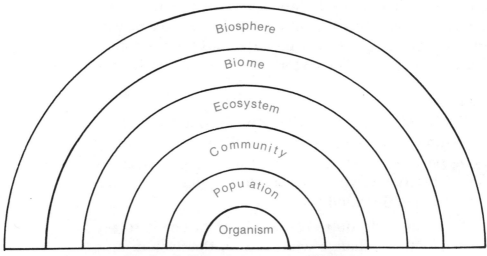

Levels of Organization in the Biosphere

climates and surroundings throughout the earth. Organisms live where they can survive, so different plants and animals live in different regions of the earth. These large regions of the earth are called **biomes**. There are groups of ecosystems within each biome. The desert biome, the ocean biome, the grassland biome, and all the biomes on earth together form the **biosphere**.

The Biosphere

The biosphere is the part of your planet where life can exist. It includes both living and non-living things. The biosphere is very large compared to you, a single organism. Compared to the entire planet Earth, it is very small. Think of the world as an orange or an onion. The biosphere is like the peel of the orange or the skin of the onion. It is the thin layer on a large sphere. It is the active part of the sphere.

All substances in the biosphere are interdependent and need each other to make life work. All known life exists in the biosphere. All of the living things are tied to other living things and their surroundings. Organisms interact with each other and with their non-living environment. *Bio* means "life." The biosphere is the area on the surface of the earth where life exists.

When you look at the earth as a whole, the biosphere seems to be very small. The thin layer can easily be damaged. Human beings are the only organisms aware of the biosphere. Humans, in a large sense, are in charge. Humans need to be aware of how the biosphere works. You need to be gentle with your part of it and take care of it. Survival of living things depends on the condition of all of its parts. The whole earth is interrelated. What happens in your part of the biosphere affects your neighbors in their part.

All life on earth is limited to a thin layer of soil, air, and water around our planet.

1. Describe the difference between a population and a community.
2. What effects do trees have on other members of the forest community?
3. Describe a habitat. Write down the names of as many kinds as you can think of.
4. Tell how a backyard is an ecosystem. Include as many species as you can think of that live there.
5. Tell what happens when a meadow is allowed to go through a natural succession. What does it finally become? What can change this?

KEY IDEA #2:

Food chains and food webs include the food and energy relationships in an ecosystem.

Food Chains and Food Webs

Links in the Food Chain

Organisms feed on other organisms in an ecosystem. A feeding pattern known as a **food chain** is formed. Big fish eat little fish, and in their turn are eaten by even bigger fish. The little fish ate water plants to begin with, or maybe fed on insects and insect larvae that fed also on plants, or protists. Actually, all food chains have their start with plants.

Plants as Producers

Green plants and some protists make their own food. They use the energy of sunlight to carry out photosynthesis. Because they do in fact produce food for themselves, they are called **producers**. They are producers of food used by other organisms in the ecosystem. Energy is the main player in any food chain, and the producers are the first link in the energy chain. Every food chain begins with some plant producer.

Consumers

Anything that does not contain chlorophyll, including all animals and animal-like protists, must get food from outside their bodies. Since plants and some green protists such as algae are the only food producers, animals must eat them. Organisms that eat other organisms are called **consumers**. Humans are among the world's consumers. Supermarkets, delis, restaurants, and sidewalk vendors all point to the piles and piles of food of all kinds that humans consume.

Why must consumers eat on a regular basis? Food fills two essential needs: l) energy to run things, and 2) materials for growth and repair of parts.

Muscle, bone, skin, teeth, hair, and blood are all products of the food you eat and the food you have eaten in the past. You and other consumer organisms must continue to eat to maintain body parts and body functions. The food consumed keeps the body parts in good working order.

Levels of Consumers

Levels of consumers are either first order, second order, or third order. It depends on when they are consumed. Eating or feeding is the biggest business of living things. **First-order consumers** eat plants. Rabbits are first-order consumers, and so are cows. **Second-order consumers** eat other animals. A hawk eats a rabbit. Each level of consumer depends on the energy stored in the body of the organism that is being taken as food.

Rabbits eat grass. Foxes eat rabbits. Foxes are second-order consumers.

The Food Pyramid

If you have seen movies of the African plains where lions live, you know that the plains are covered with grass — billions and billions of grass plants. Herds of antelopes may number in the thousands of animals, but the number is far smaller than the number of grass plants. In the same area, there may only be a few dozen lions. Lions need a wide range in which to hunt.

Any food chain starts with large numbers of producers. The number of consumers falls sharply as you go from one level to the next higher level of the chain. You may think of the food chain as a pyramid with the highest-level consumers at the top. There cannot be very many consumers at the top because it takes so many levels below them to keep them supplied with food. Yet even they are dependent on the producers.

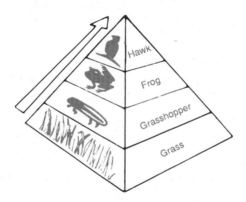

Food Webs

Very few consumers eat only one kind of food. If they did, the failure of the harvest or hunt for that one food would force them to starve. Most consumers eat a variety of foods. The American crow is an example of a bird that eats a wide range of foods.

A frog in a pond is a second-order consumer. The frog eats insects, fish, other frogs, and earthworms. The frog in turn is eaten by fish, snakes, and some birds. Frogs are

Basic Biology

part of a food web like most other consumers. A **food web** is a complex pattern of overlapping food chains. It shows the feeding interactions in an ecosystem. It shows which animals depend on which other animals for energy and for materials from their food. A food web shows that organisms depend on the producers in the community. A food chain is only a part of a food web, showing one line of the relationships or connections. Both chains and webs show how energy and materials travel from one member of the community to another.

Decomposers

Suppose that a high-level consumer like the lion lives to a ripe old age and dies in a place where it is not eaten by a scavenger. Does the food chain stop there? No, that is where the **decomposers** enter the scene.

Bacteria are everywhere, and so are molds and other fungi. Whenever an organism dies, if nothing eats it directly, it will soon be broken down, digested, and taken in by these decomposers. The job of the decomposers is to turn the food chain into a food cycle. Decomposers break down the complex materials of a dead animal or plant into simpler raw materials that can again be taken in by plants (producers) to make more food. Nothing is ever wasted.

CHECK YOUR UNDERSTANDING
1. What is the difference between a food chain and a food web?
2. Why can plants get along without animals, but animals cannot live without plants?
3. Diagram the longest food chain you can think of in the form of a pyramid. How common is the animal at the top compared to the bottom organisms?
4. What is the role of decomposers?
5. What is a food cycle?

Substances are constantly being reused in a never-ending series of cycles as living things carry out basic life activities.

Spaceship Earth

The planet Earth has been compared to a spaceship — a very large one, but a spaceship nonetheless. The earth is a planet, and we inhabit its outer surface. We use the materials we find here to build our homes, to make tools, and even to eat. If some substance is already in scarce supply, there is no way to get more of it. There are no cargo spaceships bringing gold from other planets, no trash spaceships carrying away waste. Everything that is here now, stays here. The biosphere is limited to what is here now. Like a spaceship, the earth is isolated in space.

The earth is a closed system. Energy from the sun enters, and heat energy from the earth is removed. Everything else on the earth is reused. Water to drink is the same water that was here millions of years ago. It is used, cleaned up, and used again. It is cycled and recycled. When you drink water, you may be using some water that was once drunk by George Washington, or a dinosaur, and maybe even yourself, years ago.

The Water Cycle

The constant reusing of water is called the **water cycle.** Water from puddles, oceans, lakes, and rivers evaporates and becomes a gas. When a puddle "dries up," the water evaporates and becomes part of the air. Water vapor comes from other places, too. The leaves of trees, the soil, and even your breath are sources of water vapor. The oceans are the largest source of water that evaporates into the air. Oceans are big contributors to the water cycle.

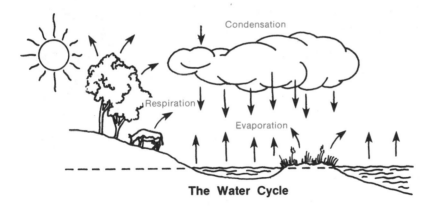

The Water Cycle

There is always water vapor in the air, but you cannot see it. Even though water vapor is invisible, you can tell that it is there. A glass with ice in it is cold, and water vapor **condenses,** or becomes liquid, on the outside of it. The water that forms on the outside of the glass comes from the air, not from inside the glass.

Warm air holds more water vapor than cold air. Warm air rises, and when it does, it gets colder. The higher parts of the atmosphere are colder than the surface air. When the temperature gets lower, the air can hold less and less moisture. The water vapor condenses and becomes water droplets. As temperatures change to colder and colder, the droplets gather closer, get bigger and heavier, and fall as rain when conditions are right. Water can be used over and over again by plants, animals, protists, and people.

The Carbon Cycle

Carbon is one of the most abundant and useful chemical elements in the world of living things. Carbon is a large percent of our bodies and of almost everything we eat. All living material is made up of chemical compounds made with carbon. Except for water and salts, all of the basic life chemicals contain carbon.

There would be no proteins, fats, or carbohydrates without carbon. All food except water and salt contains carbon.

Green plants make food using the carbon dioxide gas in the air. Protists, largely in the oceans, make a great deal of food. Animals and decomposers make carbon dioxide as a waste product and put it back into the air. Together, these producers, consumers, and decomposers constantly recycle carbon, using it over and over again.

Some carbon gets tied up in such things as fossils and fossil fuels like coal and oil. Eventually, it gets back into the cycle to be used and reused.

The Oxygen Cycle

Oxygen is one gas that is essential to almost every form of life. Plants, animals, and protists use oxygen to change their food into the energy they need to survive. Even the producers in the food chain need oxygen in order to convert the food they make into energy to maintain their own life.

All of the oxygen in the air comes from producers. Green land plants and ocean protists provide oxygen for all living things. When the sun is shining, green plants not only recycle the oxygen which they use, but are able to release more than enough of the vital gas for the consumer organisms.

The Nitrogen Cycle

Nitrogen is a gas that makes up about 78% of the air. When combined into chemical compounds, it is absolutely necessary for life. Many important compounds, especially proteins, contain nitrogen. Nitrogen also dilutes the oxygen we breathe and we do not age as rapidly.

The nitrogen in the air is not usable for making proteins directly. It must first be combined into a chemical compound such as ammonia. These chemical reactions are done by bacteria in the soil which are called **nitrifiers**. They take nitrogen from the air and change it to a form called nitrates. The nitrates in the soil are used to make fertilizer for plants. The plants absorb the nitrogen compounds and make proteins which consumers can eat.

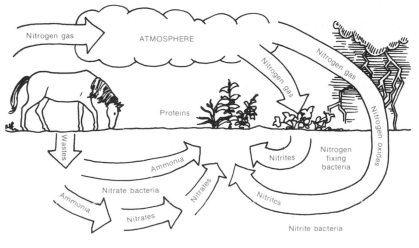

The Nitrogen Cycle

Consumers produce waste which contains other nitrogen compounds that plants can use. Once nitrogen is taken from the air by nitrifiers, it can go from plants to animals and back again in different forms. Cattle eat plants to get protein, and their manure is an excellent fertilizer for the plants.

There are other bacteria in the soil that undo the work of different nitrifiers. They are called **denitrifiers**. They break down proteins and other nitrogen compounds and return the nitrogen to the air. The amount of nitrogen in the biosphere, like other elements, stays the same: cycled and recycled, but never lost.

Cycles Work Together

None of these cycles works alone. Although a cycle can be described for each element or compound, they are really joined in nature. Nitrogen compounds also contain carbon and oxygen, and they are dissolved in the water of the water cycle. The whole ecosystem works together like a big, complicated machine. You can look at one part at a time to make it easier to study and understand. Each cycle is only a small part of the whole system. Systems are made up of working parts that do things in cycles. The cycles interact with each other.

Many substances in the ecosystem are recycled over and over again. There is a cycle for iron, calcium, phosphorus, potassium, and every chemical used by living organisms. Biology is the study of life, but it is dependent on the recycling of the non-living chemicals in the biosphere.

1. Give an example of a closed system.
2. Why does water vapor condense high in the air?
3. Why is carbon important to living organisms?
4. What organisms take nitrogen directly from the air?
5. Why does nature recycle everything? What would happen if something were not recycled?

KEY IDEA # 4

Energy is essential to all life. Energy constantly enters and leaves the earth and cannot be recycled.

What Is Energy?

Energy comes in many forms. Light and heat from the sun are energy. Electricity is also energy. Your muscles need energy to run and throw a ball. Your heart uses a lot of energy to pump blood, and your brain uses it when you think. The warmth of your body is heat energy.

Food Is Energy

Energy can be stored in chemicals like gasoline, to be released when it burns. Food is another form of stored energy.

Producers absorb energy from the sun when light falls on leaves that contain chlorophyll. Only part of the sunlight is absorbed, and only part of that is stored as food. Some of the food is used by the producer itself. Only a small part of the original energy that left the sun can be passed on to consumers.

Consumers Use Energy

The way in which consumers use energy can be compared to levels of a pyramid. The first or lowest level of the pyramid is made up of producers such as green plants. The energy stored in the plants is the greatest amount of energy at any level. The next level of organisms in the pyramid are the first-order consumers, or plant eaters. Rabbits are a good example. First-order consumers like rabbits eat producers like plants and protists. The rabbits receive much, but not all, of the energy held in the plant. Plants use some of the energy to carry out their chemical processes. The next level of the pyramid is made up of second-order consumers or animal eaters. Foxes are an example. Foxes capture and eat rabbits. The foxes receive a good deal of energy from the rabbits, but not all of it.

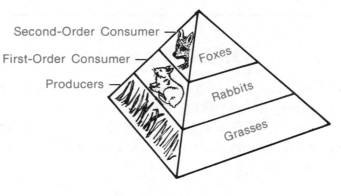

Second-Order Consumer

First-Order Consumer

Producers

Foxes

Rabbits

Grasses

An Energy Pyramid

Consumers do not produce any energy of their own. They must get all their energy from producers. It takes many producers to keep one consumer alive. A few rabbits must eat a field full of plants to have enough energy. In the same way, it takes a number of rabbits (first-order consumers) to keep just a few foxes (second-order consumers) alive. There is a loss of energy at each step of the food chain. Less of the original amount of energy is available the higher up the pyramid the consumer is from the producer.

In this pyramid, there are a great many plants at the bottom, a small number of rabbits in the middle, and just one or two foxes at the top.

Second-order consumers, like foxes, eat first-order consumers, like rabbits, to get their own food energy. However, only a small part of the food energy eaten by the rabbits is left to be passed on to the foxes. The loss of heat energy at each step of the food chain explains why there must be large amounts of energy at the bottom of the pyramid. The larger number of plants as producer organisms is shown in the form of a pyramid. Less and less of the original amount of energy from the grasses is available to each new level of consumer.

Heat

A warm-blooded organism constantly loses heat to the air which surrounds it. Waste energy ends up as heat. At each step of the food chain, less energy is stored. It becomes heat. Finally, all of the stored energy becomes heat, and there is no energy left to be recycled.

Energy cannot be recycled. And that is why the sun is so important every day. The sun sends light and heat energy back into the system.

Energy from the Sun

Producers are dependent on a constant supply of energy in the form of sunlight. If the supply were to stop, no food would be produced, and in a short time, all the stored food energy would be used.

Energy comes to earth from the sun every day, even on cloudy days. Light and heat are absorbed on the daylight side of the earth constantly. The same amount of wasted heat energy is constantly lost to space on the night side of the planet. The earth constantly takes in and loses energy at a rate that keeps it from getting hotter and hotter overall. Each spot on the earth goes through warm periods and cool periods daily and yearly, but on the average, the temperature of the whole earth remains the same.

You cannot make your own food from sunlight as the producers can. All of your food comes from the sun indirectly. You and all living organisms depend on the sun for life itself.

CHECK YOUR UNDERSTANDING

1. Why do producers need sunlight?
2. Why doesn't the earth become hotter or colder?
3. Why do moles, who live underground and eat insects, need the sun?
4. Where can energy be stored?
5. What is the difference between a producer organism and a consumer organism?

Summary of Chapter 12

Every living organism depends on other organisms for survival. The non-living parts of the environment are equally important. Together the interaction of living and non-living things make up an ecosystem.

All food comes from plant and protist producers directly or indirectly. Food energy is really energy from the sun stored by the chlorophyll in the producers. Both plants, animals, and protists depend on sunlight and chlorophyll.

All substances on earth are constantly reused in a never-ending series of cycles. Some especially important cycles are those of carbon, oxygen, nitrogen, and water. All cycles work together to keep the biosphere running.

Energy is the only thing that constantly enters and leaves the closed system of the earth. It is necessary for life and cannot be recycled.

1. *succession*
2. *ecology*
3. *environment*
4. *biosphere*
5. *pollution*
6. *population*
7. *ecosystem*
8. *community*
9. *biome*
10. *food web*
11. *decomposer*
12. *producer*
13. *consumer*
14. *food chain*
15. *first-order consumer*
16. *second-order consumer*
17. *water cycle*
18. *condense*
19. *nitrifier*
20. *denitrifier*

1. What is the biosphere?

2. Why do ecologists study pieces of an ecosystem instead of the whole thing at once?

3. With what does every food chain start?

4. What is the difference between the amounts of energy lost from each level of the food chain?

5. What does the word *photosynthesis* mean?

6. Tell what makes an environment.

7. Why is a pond community not permanent?

8. Why do many species live only in certain habitats?

9. What is ecology?

10. Name three ways people pollute their environment.

11. What do producers produce?

12. Which could survive better alone — producers or consumers? Why?

13. What two needs are filled by food?

14. Why is a food chain like a pyramid?

15. Why is the earth like a spaceship?

16. What substance is found in all of the basic life chemicals except water?

17. What do producers provide for humans in addition to food?

18. How is nitrogen from the air made into proteins?

19. Why don't we run out of water, oxygen, or any other substances in nature?

20. Why do we need a constant supply of energy?

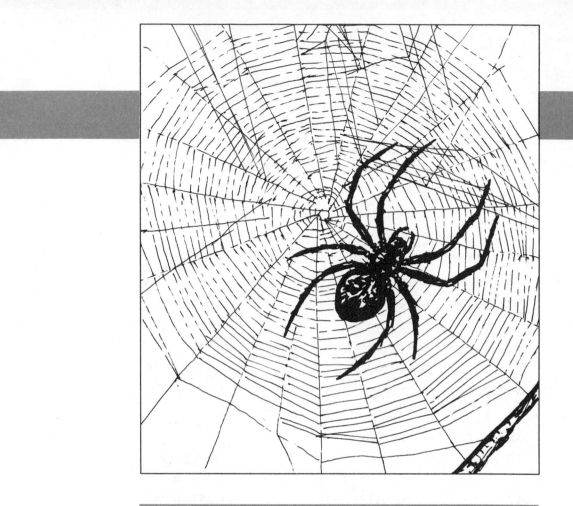

To recognize that patterns of behavior are responses to conditions in the environment.

To understand that a survival system in populations of organisms relies on patterns of behavior that include innate, learned, and communication behavior.

KEY IDEAS:

Certain behaviors of each species are innate, inborn, or instinctive

Learned or acquired behavior involves the use of thinking, reasoning, and problem solving.

Communication or signals behavior is a form of behavior that has survival value for populations of organisms.

CHAPTER 13

Describing the Behavior of Organisms

Introduction

Behavior is a term used to describe the things that organisms do. There are two main kinds of behavior: 1) those things that are done automatically and are innate, and 2) those things that are learned or acquired.

Innate or instinctive behavior is present at birth. It does not have to be learned. Heredity is responsible for built-in pathways that cause organisms to behave automatically in certain ways. Learned behavior is behavior that occurs after an animal has had some life experiences. Learned behavior is a characteristic of all vertebrate animals and is most highly practiced in humans.

Humans study behavior of animals because they are curious. They see patterns of their own behavior in the behavior of other organisms.

Certain behaviors of each species are innate, inborn, or instinctive.

Species Behavior

Innate Behavior

Innate behavior is controlled by genes that are inherited. Humans inherit a **reflex** that causes blinking when any object moves near the eyes. From the time of birth and all through life, the blinking reflex protects the eyes from harmful objects. Blinking occurs without thinking. It is a reflex act. The nerve pattern for blinking is inborn. The response is automatic.

Nest-Building Behavior

Nest-building is an inborn set of behaviors. Nest-building is an **instinct**. Birds build a wide variety of nests. They are built in different places, with different materials, and they come in all sizes. Materials for nests are a part of the habitat. A habitat is that special part of an ecosystem where each species lives its particular lifestyle. Gulls use seaweed and other seashore plant materials to build a large nest on the ground. Terns living in the same community make a circular dent in the sand where they lay their eggs. These nest-building behaviors are innate or inborn. Each kind of bird instinctively builds one kind of nest. They build it in a certain place in their habitat.

Courtship Behavior

Nest-building is just one part of a set of behaviors that are connected and inborn or innate. **Courtship behaviors** come ahead of nest-building and mating behaviors. Courtship is the set of behaviors that is used in attracting and acquiring a mate. Courtship involves a pair of organisms, but the male animal does most of the courtship display. All of the vertebrate animals exhibit courtship behaviors.

One part of the courting process is making some kind of sound. In birds it involves the song of the male. In frogs it involves the croaking of the species mating call. In a pond where several species of frogs are courting, it is important for each species to mate with its own kind. Survival of the species depends on having the right sperms with the right eggs. Otherwise proper development does not take place, and the members of the next generation are lost.

Behavior and Survival

Courting behaviors are inborn or innate. The life span of some of the simpler vertebrates is not long enough for learning the steps involved in courting behaviors. Hereditary materials provide the chemical and nervous signals for courtship behaviors. Courting behavior leads to mating. Much of this behavior must be timed perfectly. There is no room for trial and error. The survival of species depends on built-in patterns. The inherited patterns coordinate the behaviors involved in courtship, mating, and nest-building.

Incubating eggs is a part of bird instinct. Sometimes both birds share the incubation task, while other times it is the job of one or the other of the parents.

Territorial Behavior

A territory is an area that is staked out by a species of animal. A territory includes a nesting place and the surrounding ground that provides the main source of food for the species. Birds have territories that they claim by their song and by fending off intruders. The mockingbird finds the tallest of trees to perch atop and sing its territorial song. In neighboring territories other male mockingbirds sing to claim their own space.

The advantages of a territory to a species are tied to the whole life of the organism. Having an established territory is useful in getting a mating pair together. It allows them to build a home or nest and raise their young. The borders of the territory are defended, and intruders are driven away. Fighting over the territory is reduced by the fact that other organisms usually honor the "owner's" rights. Animals that are successful in establishing a territory are generally able to dominate in fights to protect their area. **Dominant behavior** establishes the leader or the winner. Winning rights to a territory aids in the process of survival.

Survival of the species is tied to the ability of a population of organisms to reproduce. Having a territory promotes the process of courtship, mating, nest-building, rearing of the young, and reproduction. The next generation that results from successful parent organisms has a genetic makeup that favors future success. **Territorial behavior** insures a good food supply for a family, an uninterrupted breeding pattern, and reduction of fighting and killing.

Imprinting Behavior

Newly hatched geese are called **goslings**. Ordinarily they attach themselves to the mother goose. In experiments to study behavior, the eggs of geese were removed from the nest and hatching took place in front of the experimenter. The baby goslings attached themselves to the first object they saw. In one case the experimenter was Konrad Lorenz, a famous biologist who studied behavior of animals. He described this social behavior of the goslings and called it imprinting. Goslings follow their "first sight" leader whether it is their natural mother or whether it is a person or even an object. Lorenz found out that once imprinting behavior gets set, it does not change. His experimental geese followed him everywhere. He became their "parent."

Goslings attached to a hen instead of their own mother.

Imprinting is "instant learning" and is of survival value to the species. Goslings in the wild become a fixed part of a family unit. This behavior plays an important part in the life cycle. By keeping the goslings in the family circle, it encourages courtship, nest-building, mating, and rearing of the young.

Spider and Bee Instincts

Instinctive behavior is not limited to birds. Spiders make and use silk for building their webs and lining their nests. Weaving a web is a complex set of behaviors that each species of spider inherits. Newly-hatched spiders do not get a lot of time with their parents. They are on their own quickly. They could not survive if they had to go through a long web-building learning process. Building a web is important in capturing food. Wrapping a captured insect in silk would be difficult to learn. So web-building and insect-wrapping depend entirely on the instincts of the spider. Without food, no spiders!

The same is true for bees. The life of the bee is one continuous set of instincts. Building the hive, finding and getting food, caring for the queen bee, protecting the hive, and swarming, are all inborn behaviors. There is no time in the life span of the bees for learning. Bees must do all of these things as a result of inborn behavior patterns.

Bees build six-sided cells.

Plant "Behavior"

Behavior is usually thought of only in connection with animals. Behaviors are observable changes. Plants do make some changes in response to stimuli. Light, dark, gravity, touch, and chemicals cause "behavior" in plants. Parts of plants turn toward the light, roots grow toward the center of gravity, and flowers and leaves open and close.

Phototropism

Potted geraniums placed on the window sill will show the broad part of the leaves facing the source of sunlight. When the plants are turned around so that the broad side of the leaves face the room, it takes just hours before the plants turn back around. The sunlight causes chemical changes that make cells on one side of the plant get longer. The plant is forced to bend toward the light as the longer cells on the outside act against the shorter cells on the inside. The tug-of-war turns into a push of war. The longer cells are successful!

Daylight has an effect on the opening of morning glories. The flowers open to the fullest extent, and then, as evening approaches, the flowers curl closed for the night. The reaction of plants to light is called **phototropism**.

Geotropism in Plants

Radish seeds and bean seeds planted in the laboratory showed that roots grow "down." Growing "down" is a response to gravity and the behavior is called **geotropism**. What would be your guess about which way the stem of plants would grow?

Response to Touch

The closing of a Venus's flytrap is in response to the presence of an insect landing on the sensitive "hairs" that line the surface of the leaves. The pressure of an insect on the fine "hair" causes pressure changes inside the plant. The leaf closes and captures the insect.

Another touch response occurs in the mimosa, or "sensitive plant." Rows of leaflets, usually extended are brought together by the pressure of touch. The closing happens in a matter of seconds. The twining of vines around tree trunks, fences, posts, and trellises is a "behavior" in response to touch or contact with the object being wrapped around.

A twining vine

A major feature of the cells of plants and animals is that they are sensitive to changes in their surroundings. Individual cells like ameba and paramecium react to chemicals and objects in the environment. The stuff that cells are made of is often called cytoplasm. This material is unique among all of the materials of the world. One of its main features is that it reacts or responds to stimuli. It is in this active and responsive material that the foundation or basis for the life activities is located. You have probably heard the word sensitive used in relation to a "sensitive person" — a person who is responsive to the actions of other people. Cells and tissues of organisms are sensitive. They respond to signals from objects and events in the environment.

CHECK YOUR UNDERSTANDING

1. Describe a main feature of an innate behavior.
2. What is the value of territorial behavior in animals?
3. What is imprinting behavior?
4. What is the value of instincts in spiders and bees?
5. What is similar about animal and plant behavior?

KEY IDEA #2:

Learned or acquired behavior involves the use of thinking, reasoning, and problem-solving.

Learned Behavior

Conditioned Response

An aquarium full of goldfish or guppies has a regular feeding time of 5:00 P.M. As the keeper approaches the aquarium at 5:00 P.M., the fish move to the top and to one corner of the aquarium. The keeper taps on the rim with the container holding the fish food. The fish scurry about at the surface. The behavior of the fish is a type of **learned behavior**. Their movement to the surface is a **response** to the **stimuli** of the 5:00 P.M. approach of the keeper and to the tapping sound of the food container. The food itself is the primary stimuli. What do you think would happen at other times during the day if the keeper simply walked over to observe the fish? If you said that they would just go about their business as usual, you would be right! The behavior pattern of the fish is part of a learned behavior called a **conditioned response**. The fish have become conditioned to the 5:00 P.M. feeding time, the approach of their keeper, and to the tapping sound of the container followed by the presence of food.

Pavlov's Dogs

Learning about what behavior is has taken some complicated behavior on the part of scientists. Ivan Pavlov, a Russian scientist, was interested in studying the digestive process and feeding behavior of dogs. His work describes the kind of learning known as conditioned response. Production of saliva in the mouth is a reflex action. Dogs salivate a great deal. They produce saliva at the sight and smell of meat, their preferred food. Dogs swallow whole pieces of meat, and the flow of saliva helps that happen. Saliva helps the food get down the gullet to the stomach.

Pavlov figured out a way to collect the saliva and measure the amount of flow. He showed the dogs meat, and collected the saliva. When the dogs were accustomed to the practice, Pavlov began ringing a buzzer at the time he presented the meat to the dog. After a number of repeated buzzer and food servings at the same time, Pavlov then rang the buzzer by itself. The dogs produced saliva in amounts the same as when they received meat. The sound of the buzzer alone caused the dog to show the same behavior as the natural sight of meat.

Pavlov found that if he continued to press the buzzer without offering the food at the same time, the dogs' flow of saliva was reduced. After repeated use of the buzzer with no food, the saliva flow was reduced to nothing. The dogs had learned to associate the buzzer with food. The natural flow of saliva at the sight of good food is innate behavior. Response to the buzzer and food was a conditioned response. Conditioned responses are a learned behavior. Learned behaviors can be "unlearned." So it was with "the buzzer and no food" response. Pavlov earned the Nobel Prize for his work on digestion.

Skinner's Rats

B. F. Skinner, an American scientist, used the rat, *Rattus rattus*, as the subject of his experiments on behavior. It turns out that rats are always hungry. Skinner designed a box in which a rat could press a lever, and a food pellet would appear. The rat, following its pattern of always being hungry, continued to press the lever. Every time the lever was pressed, it was automatically recorded. Skinner found this went on for hour after hour. Just as in Pavlov's dogs, if the food supply is cut off, the rat stops pressing the lever after a while.

Skinner went a step further by giving food when the lever was pressed, but only when a light was on in the box. In the dark, the rat did not press the lever bar. The rat learned the difference between "food in the light" and "no food in the dark." The rat had figured some things out. The rat found out that its behavior caused something to happen. The rat learned that the light made a difference.

Adaptation

The key difference between the dog experiment and the rat experiment is that the dog response was a reflex with saliva flowing just before the food was taken. In the rat experiment, the food appeared after the lever was pressed. Learning that food would appear with the pressing of a lever led to success in food-getting for the rat. The rat had adapted to a set of conditions in the environment. Adaptations in natural populations of animals lead to similar successes. Successful adaptations are helpful to the continuation of the species. Species are able to continue if a population can adapt to change in the environment.

Bird Song Behavior

Is the song of the mockingbird or the chaffinch a learned behavior — or is it inborn? Research with birds raised in isolation indicates that a basic birdsong pattern is inherited. Birds hatched from eggs taken from the nest and brought into the lab were able to sing the species song. But, they did not sing the elaborate, or full, song of the parent birds. When placed in the presence of the parent birds, some learning takes place by imitation. Birds and primates have been the major animals used in the study of behavior.

Reasoning

The key to learned behavior is that it depends on the ability to reason. Reasoning involves making connections between ideas. Learning involves memory, which is the ability to recall or reproduce ideas. How memory works is not completely understood. Reasoning involves the use of **logical patterns** of thinking. The end result of learned behavior is the ability to solve problems.

Chimpanzee Behavior

A biologist placed a chimpanzee in a room to find out what it might do. The scientist peeped through the keyhole. What do you think he saw? Yes, the eye of the chimpanzee was looking out of the keyhole from the other side. The chimp was just as curious as the scientist!

The same scientist put a banana on a string and hung it from the ceiling. The banana was out of reach of the chimpanzee. Several boxes were placed on the floor around the room. When the scientist looked through an observation opening in the door, he saw that the chimpanzee had the banana. The chimpanzee had solved the problem by putting the boxes on top of each other, climbing to the top, and easily reaching the banana. The chimp had put some ideas together. The problem involved use of the brain. Reasoning took place.

Problem-Solving

Problem-solving is the highest kind of learned behavior. Living things are limited in the kinds of behavior they can use by the kind of nervous system they have. Learning behaviors are features of more highly developed vertebrates. The ability to learn and to solve problems governs the lifestyle and patterns of activity of organisms. Human beings have the most complex behavior patterns of all living things. Humans are the best problem-solvers. They are the best at solving problems because they have developed the highest system of communication. Complex and useful language patterns separate humans from all other organisms.

1. What conditioned response did Pavlov discover?
2. Describe a major feature of learned behavior.
3. Tell about the behavior of chimpanzees.
4. What is the highest kind of learned behavior?
5. What separates humans from all other organisms?

KEY IDEA #3 :

Communication or signals behavior is a form of behavior that has survival value for populations of organisms.

Communication Behavior

What Is Communication?

Communication is an activity of an organism that uses some form of signal or stimulus. Communication is behavior. Communication may be a signal from an organism to another of its kind. The signal may be sent to other kinds of organisms in the environment. Most communication behaviors take place between members of the same species.

Bees have a system of communication related to finding food. They do a dance pattern that indicates to other bees the direction and distance of flowers in their environment. Knowing where to find food is important to the survival of the hive and the species.

Birds have elaborate communication systems for getting and keeping a mate. They communicate through song to keep their own territory. Baby birds open their mouths to be fed. The beak of the parent bird communicates by markings. Male birds get into fighting postures with other males. They gain or keep their dominant place in the flock by such behavior. Most birds have a communication signal that causes the group to take flight at some sign of danger to the flock.

Sign Language

Humans have tried to teach chimpanzees and other apes how to use the **sign language** of the deaf. Language is a part of behavior. Animals respond to the stimulus of the signs and sounds of communications behaviors. A recent news article told of a chimpanzee that uses "hundreds of sign language words." A baby chimp learned 55 sign words from an adult chimpanzee.

Also reported was a story of an orangutan that had used sign language to "ask to be taken for a car ride," "took along money it had earned for keeping its room clean," and "gave the driver directions to the local Dairy Queen."

Chimpanzees Teach Sign Language

Scientists Told That Apes in Experiments Instruct Each Other

Primates have the highest level of behavior among all of the animals. Among the primates, human beings have the highest level of intelligent behavior.

Human Behavior

The key idea in human behavior is that human beings are reasoning organisms. The ability to think about everything and to remember an almost endless number of words, ideas, and events is remarkable behavior. Humans are problem solvers, judgment makers, and decision makers.

Humans establish governments and religions, create states, and make laws. Humans run on the land, swim in the seas, fly through the skies, travel in space, wander on the ocean floors, climb the highest mountains, and make their homes in every kind of place. Humans build and run large factories and large cities.

Humans also pollute the air, water, and land. They foul the atmosphere with hydrocarbons from their automobiles. They have altered the atmosphere in ways that affect global temperature conditions. This alteration can, over long periods of time, affect the vegetation, climate, and living conditions of all species. One famous writer claimed that "humans are the only animals that blush — and the only animals that need to!"

Fortunately the human species is a problem-solving and decision-making organism. Problems that have been created are able to be solved. Modern technology and communications systems have improved the opportunities to do so.

Human Culture

All these behaviors are based on the human ability to make and use language to communicate. Because of the written and spoken word, the collective **human culture**—all of knowledge and history—is able to be passed from generation to generation. New populations benefit by having a storehouse of information and ways of doing things that have already worked. New knowledge is added with each generation.

This **cultural evolution** is strictly a human event. *Homo sapiens* are now a part of both the process of biological evolution and cultural evolution. Changes in human culture are happening faster than changes in biological evolution. Humans are more in charge of events that affect the human species. Humans alter the environment to improve their chances for survival.

Advantages of a Long "Growing-Up" Period

Humans have the longest growing-up period of any organism. Humans are infants and children for years. They have an adolescent "growing up" period and even a long young adult period when they have a chance to learn what is needed to get along in a complex world.

Human communications and other kinds of learned behaviors continue all through the life span. Human beings solve problems, gain new knowledge, add it to prior knowledge, and use it to solve new problems. Human behavior is affected by previous behavior. Humans often say that "experience is the great teacher" and so it is!

CHECK YOUR UNDERSTANDING

1. How do bees communicate with other bees?
2. In what way have chimpanzees and orangutans been taught to communicate like human beings?
3. Describe how human behavior is different from the behavior of other animals.
4. How have humans affected this planet?
5. What is the advantage of a long growing-up period for humans?

Summary of Chapter 13

The two main types of behavior are innate behavior and learned behavior. Innate behavior is inborn or instinctive. Heredity is responsible for this behavior. An organism is born with this type of behavior. Nest-building, courtship patterns, and displays are examples of innate behaviors in birds. Weaving a web is an instinctive behavior of spiders. Some animals are born with imprinting behavior. Even plants instinctively respond to stimuli. Innate behavior helps a species survive.

More highly developed organisms, especially vertebrates, can also learn things. Learned behavior is acquired in different ways, such as by conditioned response and by trial and error. The highest form of learned behavior is problem- solving. Humans are the best problem-solvers.

All animals have a system of communication. Bees communicate by dancing. Birds use song. But humans can communicate through speaking, reading, and writing. In this way, people can store information and pass it down from one generation to the next.

1. *behavior*
2. *innate*
3. *reflex*
4. *courtship behavior*
5. *goslings*
6. *territorial behavior*
7. *dominant behavior*
8. *imprinting*
9. *phototropism*
10. *geotropism*
11. *instinct*
12. *learned behavior*
13. *stimulus*
14. *response*
15. *conditioned*
16. *logical patterns*
17. *communications*
18. *sign language*
19. *cultural evolution*
20. *human culture*

1. What is an instinct?

2. Define a habitat.

3. List reasons for courtship behavior in birds.

4. What is the advantage of courtship behavior being inborn instead of learned?

5. What is territorial behavior?

6. How do courting behaviors, signals, and territorial behaviors help a species survive?

7. What is the value of imprinting?

8. What innate behavior does a spider have?

9. Define phototropism.

10. Define geotropism.

11. What is the difference between innate and learned behavior?

12. What is a conditioned response?

13. How did Pavlov condition his dogs?

14. What was the difference between the behavior of the animals in Pavlov's dog experiments and in Skinner's rat experiments?

15. Describe the highest kind of learned behavior.

16. What are some methods of animal communication?

17. What are some methods of human communication?

18. List ways that humans endanger the planet Earth.

19. Describe the cultural evolution of humans.

20. How do humans use communication behaviors to survive?

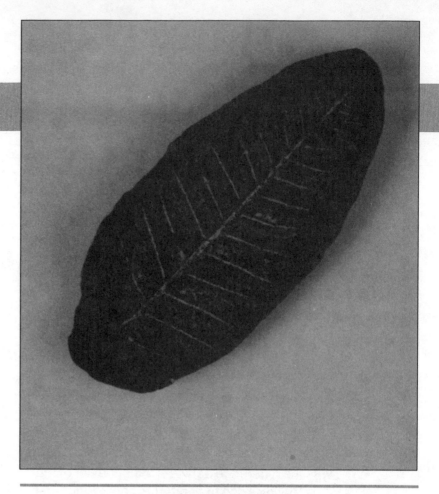

To trace the fossil record as a means to understanding the history of life on earth.

To describe the main ideas of the theory of evolution and natural selection.

KEY IDEAS:

Fossils provide evidence for the existence of different kinds of organisms over long periods of time.

The theory of evolution and natural selection offers explanations for changes in natural populations of living things.

Since its beginnings in the vertebrate fossil record, the human species has become most widespread among the higher organisms.

CHAPTER 14

Tracking Changes In Organisms Through Time: Evolution

KEY IDEA #1

Fossils provide evidence for the existence of different kinds of organisms over long periods of time.

Fossil Evidence

What Are Fossils?

Fossils are the remains, impressions, or other traces of animals and plants of the past. Fossils are preserved in the earth's crust. The word "fossil" comes from the Latin word *fossilis*, which means "dug up." Different kinds of fossils are discovered in the different layers of rock. Fossils are uncovered as scientists dig.

Paleontologists and **anthropologists**—scientists that study the past—dig for and study fossils. Humans are very curious about life in the past. The history of life on earth is a long one; that history is called the fossil record.

The Fossil Record

The 3-billion-year history of life on earth can be studied in the fossils that have been found in almost every part of the world. Any trace of plants or animals from the past is considered a fossil. **Fossil fuels** in the form of coal, oil, and natural gas are products formed from once-living things. Other fossils have been found enclosed in **amber**, a yellowish mineral that is transparent. This kind of fossil is called a **mineralized fossil**. Tracks and footprints found in rock are fossils. Other fossil records are in the form of **petrified wood**. Over long periods, chemical changes have replaced the wood with rock. But the original form and structure of the parts of the woody tree have been preserved — a history of plants made in rock.

Another interesting fossil record is in preserved animal dung, or solid waste. This fossil record is of special value because it tells the story of types of food eaten by the animals of the past. The story of plant life follows along with the story of animals.

Fossils and Evolution

Fossils provide information about the structure of animals and plants from all parts of the earth. The fossil evidence gives indirect support to the theory that there have been changes in animals and plants over very long periods of time. Fossils found in layers of rock next to each other are more alike than fossils from layers far apart.

The horse has been traced in the fossil record through 60 million years. The teeth of horses and their hooves show a series of changes that indicate **adaptations** to changing conditions. Even the condition of the surfaces of the teeth show different kinds of wear and tear. The horse of 60 million years ago was much smaller than modern-day horses. The fossil record shows this horse was less than twenty inches high.

Fossil Variation

The fossil record shows sea-life fossils from as long as 500 million years ago. In the earliest fossil records there was not as much variety of life forms as in the more recent records. The records indicate an increase in the different kinds of life with the passing of time. Fossils from different major groups of animals show up in rock formations. An "age of fishes" preceded the "age of amphibians." In each "age" that form of life was the dominant one.

The age of the earth has been estimated to be 4.5 billion years. **Geological timetables** show the first fish fossils from about 500 million years ago, amphibian fossils from about 350 million years ago, and reptile, insect, and fern fossils from about 200 million years ago.

Dating Rocks and Fossils

Finding out the age of a certain rock formation takes some detective work. If you know something that is a fact, then you can build other ideas around what you know. Scientists know that certain minerals are **radioactive**: the mineral gives off rays, and it changes to another substance at the same time. Uranium is a radioactive element that gives off rays and changes to lead. The change is very slow, but very regular. In fact, the rate of change of uranium to lead is like a clock ticking off the seconds, minutes, and hours. But instead of seconds and minutes, the "radioactive clock" has ticked off and measured years.

Uranium has a **half-life** of 4.5 billion years. The half-life is the amount of time that it takes for a chunk of uranium to become one-half lead. Because the rate of change is constant, the amount of uranium-to-lead comparison is just like reading the hands on a clock. The main difference is that the measurement is in calendar years instead of minutes. Fossils are dated by knowing the age of the rock formation in which they are found. Carbon 14 and strontium 90 are also radioactive elements. They are frequently used for the process of dating fossils.

The Saber-Toothed Tiger

The extinct saber-toothed tigers first showed up in the fossil records of about 40 million years ago. The fossil remains were found in a rock formation where the rocks contain radioactive minerals and are able to be dated. The saber-toothed tigers became extinct only about 20,000 to 30,000 years ago.

"Ancient Wing"

One of the famous fossil finds was that of a short-winged creature with feathered wings. This fossil also had a lizard-like tail and a beak with teeth like those of a reptile. The evidence shows an organism from 140 million years ago that had features of both reptiles and birds. This organism, called *Archaeopteryx*, got its name from the Greek for "ancient wing." It was a wobbly flier because of its short wings. It had claws at the front of both wings, probably for catching hold as it made clumsy landings in trees.

Fossil Animal Connections

What is the relationship of animals of the past to animals of the present? There are connections that can be made. The fossil record shows forms of life in the past that are similar to forms living today. Scientists base information about fossil forms on what they know about the animals of today. On that basis, they are able to look at fossil bones and teeth and reconstruct the rest of the animal body. From actual measurements and direct observations, the scientists are able to put two and two together to make indirect observations about the fossils.

Fish fossils have been found in sedimentary rock layers from 530 million years ago. There are records of fish fossils in many different layers of rock. Their history stretched down through the years to the present living species. Most of the record is of impressions of bones made in the rock. Reconstruction of the whole animal is based on what is known about fish that are living today and thought to be relatives of those early fish. In addition to the kind of fish, the fossil record tells something of the number of kinds of fish. The record indicates that there is a greater variety of fishes today than at any time in the lengthy fossil record.

Coelacanth fish

Some "living fossils" are the horseshoe crab, opossum, and the coelacanth fish. They are alive today, and also show up in the fossil record. Horses and spiders have also been around in some related form for millions and millions of years.

To show some of the relationships of life forms from the past to animals of the present, the following description of a "model tree" is used. The main trunk is anchored firmly in the ground of the past and shows the beginning of the fossil record. The trunk is wide because it is old (4.5 billion years), but it shows only a small number of different kinds of organisms. The fossil record is slim. However, the fossils that are there represent the oldest record of life on earth. The first fossils do not show the first life on earth. They do show the first life with hard parts like shells and bones. It is certain that simpler forms existed before the first fossil records, but they did not leave fossil records.

As you go up the trunk to the first branches of this imaginary tree, there is an increase in the variety of life forms. Ancient clams, corals, sponges, and a large number of other invertebrate organisms have left parts of their remains in the rock record.

The most ancient fossil record is thought to be 3 billion years old. There are skimpy remains of sponge skeletons, some worm holes, and some shells of marine microscopic organisms. Paleontologists continue to work on the problem of what appeared in the fossil record and when. The work is not complete, but the record becomes clearer every year. Relationships between fossil organisms and living things on earth today continue to be uncovered.

Hunting Fossils Today

Recent expeditions in Thailand have traced the development of vertebrate animals. The fossils have been dated as being from between 240 million years ago and 65 million years ago. This is commonly known as the "age of reptiles."

Fossils found in the sedimentary layers show that when upheavals formed mountains, the sea was forced to retreat. Then a fresh-water lake was formed. The most abundant animal fossils were those of fishes. Scales of old species were identified in the rock history, as were toothplates

Lungfish

from lungfish. These toothplates, which are formed from fused teeth that were used to crush food, tell of air-breathing fish. There are living lungfish in Africa and Australia. Remains of heavily armored reptiles were recognized from their bones and teeth. These reptiles are related to crocodiles, which appear later in the fossil record.

CHECK YOUR UNDERSTANDING

1. What are fossils?
2. In what kind of rock are most fossils found?
3. Describe the origin of fossil fuels.
4. List two kinds of fossils other than the ones found in sedimentary rock.
5. What does the fossil record show about the teeth and hooves of horses?
6. Tell the meaning of the "age of fishes" or "age of reptiles."
7. What is the main idea behind radioactive dating?
8. Describe a method for telling the age of the now-extinct saber-toothed tiger.
9. Tell the meaning of the term "living fossil."
10. How old is the oldest fossil record?

The theory of evolution and natural selection offers explanations for changes in natural populations of living things.

Evolution Is a Process of Change

Organisms live in populations. Any population is made up of individuals of the same species. Species are made up of individuals of the same kind who are able to interbreed and reproduce. **Evolution** is a process of change that occurs through the individuals in a population.

Populations of organisms continue to exist over long periods of time—much longer than any of the individuals in the population. The process of evolution takes place over generations. Each new generation is the product of the genes and chromosomes of the current generation. The theory of evolution describes the results of genetic changes in organisms over periods of time that have led to the formation of new species. The theory also offers explanations for the relationship of living things to each other and to fossil remains over long periods of time—millions of years.

Populations, Genes, and Earth Changes

Evolution is a process that involves the change of the genes in a population of living things. The genes are the determiners of traits and characteristics of the individual organisms in the population. If the genes are changed or altered, that affects the traits and characteristics. Radiation, ultraviolet light, and chemicals in the environment may act on genes and cause them to change the organism. If the new gene changes are useful to the organism, they will improve chances for survival. The traits for improved survival are passed on to the next generation by way of genes and reproduction. If traits are harmful to the organism, it usually dies before it reproduces.

If the environment changes suddenly, and the conditions for living are not able to be met by a population or group of organisms, the population or group will die out. The fossil record indicates that such changes have taken place. For example, the dinosaurs were once a dominant group of reptiles that are now extinct. The group of dinosaurs had a pool of genes that influenced the traits that allowed them to survive for millions of years. Whatever the reason for their extinction, they once survived for a long period of time by changing to forms that were able to survive in changing environments.

Splitting a Population

An interbreeding group of animals may be separated into distinct groups by some major change in the earth. A large river or mountain barrier may form so that one group is on each side. The species is now forced to live apart. They do so for millions of years. The group on each side is now made up of a separate gene pool. Generation after generation, there are changes in the genes. Changes also occur in the ways that the genes are grouped and regrouped through sexual reproduction. The changes have to do with different ways of getting food or changes in mating behaviors brought on by chemical changes. Eventually the separate populations become so different that if the occasion arose where the two groups came together, they would no longer be able to interbreed. New species have been formed.

Another example of splitting a population takes place where there is an island offshore from a mainland population. Animals from the mainland travel or are transported across the water. On the island, they interbreed, and reproduce for generation after generation. Changes occur. Competition for food and space takes place. Natural changes in the genes occur in the population. The original population on the island becomes more and more different. New and different populations arise. Different species may result.

Organisms Share Common Traits

The genetic material of all organisms is made up of two chemical substances called DNA and RNA. DNA and RNA are called **nucleic acids**. The nucleic acids are all made of the same basic chemicals. This strong sameness of chemical makeup in all living things indicates a relationship. Creatures of the past show major similarities with animals of the present.

Recent studies of changes that occur in genes of organisms have shown that there is a change in DNA. This information makes it possible to make estimates about changes in animals and plants and compare it with very precise dating of fossils. Protein changes that take place because of mutations do so at regular rates. Protein changes are related to DNA changes. The DNA changes and the process of radioactive decay used to date fossils in sedimentary rock formations both help to explain evolution.

More and more evidence is becoming available to show the relationships between organisms that are alive today and those that appear in the fossil record. The fact that these relationships exist is a big part of the theory of evolution.

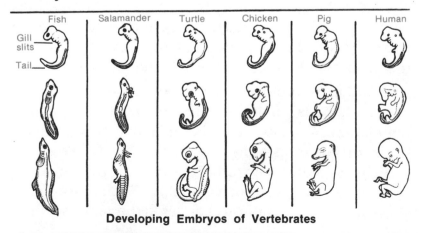

Developing Embryos of Vertebrates

Organisms in general make use of similar enzymes. They make use of chemical energy in similar ways, and they go through patterns with similar stages in their early growth and development. Development in some embryos show stages that resemble a history of the evolution of the species. Mammal embryos have a stage where the embryo is very fishlike. The basic structures of animals found in the fossil record are similar to structures of modern day organisms.

Darwin and Fossils

Charles Darwin was a student of many subjects. He studied **geology** (science of the earth) and checked series of fossils that were from rock formations that were known to be very old. He compared them with fossils from younger rock formations. Darwin found similar organisms in both sets of rock formations and noted that similar organisms had undergone change. The question being asked at the time was "Do species ever change?" The fossil evidence showed that species did change, and that indeed, new species could be formed.

Darwin offered explanations for how the changes might have taken place in populations of organisms. He made detailed observations and studies of changing rock formations in his native country. He made estimates of how long it might have taken for a well-known land formation to come into being. There were different fossils of similar organisms in different layers, and they were much older in some layers than in others.

Charles Darwin

When all of Darwin's many observations were put together, he had described a reasonable and logical set of conditions about the way that many natural events took place. This is what led him to write and publish his famous book, *On the Origin of Species,* in 1859.

Darwin and Natural Selection

A key idea in Darwin's work was that the conditions in nature, or natural situations, have an influence on which organisms in a population will survive and reproduce. This process of natural selection is what results in evolution of a species. **Natural selection** takes place as a population produces more offspring than will survive. In populations with very large numbers, there are greater chances of a variety of traits. This greater variation within the population results in increased numbers of individuals better suited to survive. These better adapted individuals are the ones that live to reproduce. The offspring are also better adapted to survive because their parents were able to. Over long periods of time, there are more chances for changes that lead to new kinds of organisms. Each generation has a genetic makeup that equips it to meet the challenges of conditions in the environment. As the environment changes, the organisms are affected. Those organisms in the population that are best suited with traits that meet the changes in successful ways will survive. They will live to reproduce the next generation.

Natural Selection

Young snakes have a tooth located in their upper jaw that is used to cut their way out of the egg shell. This temporary tooth is the only means of getting out of the shell. Snakes that are able to do this with ease are better prepared. The selection of young snakes within a population that are successful at breaking out of the egg helps promote the continuation of the species. Unsuccessful young snakes obviously do not live to breed, and their genes are lost from the population. The presence and use of the **egg tooth** is an example of natural selection acting in a population. The absence of a useful egg tooth eliminates certain snakes from the population.

Young birds still in the egg also develop an "egg tooth" to help break the eggshell. The behavior is innate. The structure of the egg tooth and its appearance at the time near hatching is controlled by heredity.

Natural Selection at Work

This is a true story about natural selection at work. In Great Britain before the industrial factories caused blackening of tree trunks, there were normal and regular populations of *Biston betularia,* the peppered moth. Observations of early collections of these insects showed some light and some dark moths, but mostly lighter moths. In the years after factories were in operation, with their smokestacks spilling out carbon soot, the tree trunks had become dark. Collections of the peppered moth showed a great increase in the number and percentage of dark-colored moths. If you were guessing about why this happened, what would you say?

If you guessed that natural selection was at work, you are right! Moths are excellent food for birds. When the tree trunks were lighter, before the factories, any dark moths were picked off by the birds because they were easier to see than the lighter colored peppered moths. As the trees got darker, the situation changed!

Now the light-colored peppered moths showed up better on the dark background of the tree trunks. Birds spotted them more easily. Then there were fewer to be collected by the English naturalists.

The environment favored the darker moths. They survived to reproduce in larger numbers. They became the dominant type of moth in that area. The darker-colored moths against the darker background kept birds from picking them off. The lighter-colored moths were captured more readily and more often. Their numbers in the population decreased.

A Population of Mice

A story similar to the moth story can be told about a population of white and dark mice living on a white sandy beach. The white mice became a larger part of the total population. They survived at higher rates. They blended into their background and were less often captured for food.

A change in the environment took place. A black flow of lava poured down on parts of the beach. The invasion of the lava now favored the darker-colored mice in the beach population, and soon there were more dark-colored mice. The white mice were now more visible on the dark lava flow. They were now picked off more often. The white beach that favored them with a better environment for survival had changed to black.

The color of mice is controlled by genes and heredity. With more dark-colored mice surviving over each generation, what can you say about the future generations in that environment?

If you said there would continue to be more dark-colored mice, you are right! This change in the nature of the mice population is explained by the theory of natural selection. The theory of the origin of the species by natural selection proposed by Charles Darwin makes use of this kind of example to explain how it works. Other workers before him had made similar suggestions, but Darwin was the first person to do the extensive travel and hard study that was needed to formulate the theory.

Darwin's Finches

When Darwin visited the **Galapagos Islands** off the coast of Ecuador in South America, he was impressed with the variety and distribution of living things. On this set of islands, named for the Spanish word for tortoise, he found a variety of birds called finches. There were also finches on the mainland of South America. He observed fourteen different species of finches on the islands. Darwin noted that they resembled the species on the mainland. They were different, however. There were differences in the size and shape of the bills. These birds are still known as **"Darwin's finches."**

Darwin recorded the fact that the shape of the bill and the size of the bill was different in each of the fourteen island species. The differences were related to differences in the diet of the birds. The birds also stayed in different parts of their island environment—some on the ground and others in the trees.

Darwin reasoned that the island species were probably descended from the mainland species. Separated by the water barrier for a long period of time, members of the finch population had undergone genetic changes that influenced the bill. Changes resulted in a bill structure that favored a part of the population, but only in a special part of the island environment.

Success at "food-getting" is a major survival trait. As each kind of finch developed and changed, it was better suited to eating a different kind of food item. One kind of finch favored eating small insects, while another finch lived on small cactus seeds. Another finch fed on large seeds and nuts, which it crushed in its heavier and larger beak. A closely-related finch fed totally on buds and fruits. One finch species fed on the larvae of insects.

A highly special adaptation to feeding is carried out by one of the finches. A cactus spine is used to rout out insects that live under the bark of trees. This "tool-using finch" reminds us of the common woodpecker. A woodpecker uses its highly-specialized beak and tongue to reach under the bark of trees for its insect food.

1. How are populations or groups of organisms a part of the process of evolution?

2. What part did dinosaurs play in the process of evolution?

3. Describe what happens when a population of organisms is split into two separate groups.

4. What role do islands play in the evolution process?

5. Describe the relationship between organisms of today and those of the past.

6. What are some ways that organisms in general are alike?

7. What did Charles Darwin's study of geology have to do with his ideas about the theory of evolution?

8. Give the title of the book that Darwin wrote and published in 1859.

9. How does "natural selection" work? (Use the egg tooth of snakes and birds to explain.)

10. Describe how dark-colored peppered moths were favored in the area of the soot-covered tree trunks.

KEY IDEA #3

Since its beginnings in the vertebrate fossil record, the human species has become most widespread among the higher organisms.

Fossil Humans

Homo sapiens is the only primate alive today that walks upright on two legs. The earliest fossil record of an animal that walked upright is 3.8 million years old. This fossil, named Lucy, was found in 1974. It shows a skull with an opening for the spinal cord at the bottom of the skull. This is an indication of walking upright. It also shows a pelvis that supports the concept of walking upright on two legs.

In the 1950s, Mary and Louis Leakey, famous fossil hunters, discovered fossil humans that they named *Homo habilis,* not *Homo sapiens.* The brain size was estimated at just more than half the capacity of *Homo sapiens,* 700 cubic centimeters. The Leakeys found simple tools at the site and estimated the age to be about 1.5 million years.

The most recent fossil finds of humans include two groups. The **Neanderthal** has been dated in the last 100,000 years. It has a brain capacity near that of *Homo sapiens* at 1300 cubic centimeters. The other fossil find is that of **Cro-Magnon.** Cro-Magnons existed about 50,000 years ago along the coast of France. They had fine weapons. They made beautiful cave drawings of the animals that they hunted. Cro-Magnons probably replaced the Neanderthals.

Cro-Magnons are described as modern humans by their genus and species: *Homo sapiens.* They share in the characteristics of modern humans: 1) the largest brains of all primates, 2) walking erect, 3) opposable thumbs, 4) tool-makers, and 5) use of art symbols.

Art found in the cave of Cro-Magnons

Homo Sapiens Today

Modern-day humans have a complex language that uses symbols for words. Humans are able to put ideas together and to make written records. Humans have a culture that is based on words that are written and spoken. The culture is information that is for human use, about humans, and about everything that has been gained and accumulated as knowledge.

Modern humans have the greatest brain size and capacity of all the primates. With their upright posture, their opposable thumbs, their large brain, and with their ability to store information, humans are able to make and use tools for all purposes. Tool-making is the highest order skill.

Humans and the Planet Earth

The human species is the only species that has the brain capacity to think about the earth. Humans are, in a sense, in charge of the planet. They know that it is a closed system, and that resources in the earth itself and its waters are limited in supply.

Having studied the science of biology, you have a better knowledge of the many relationships that exist among the living and the non-living materials of the earth. You have a better understanding of the fight to stay healthy, and of the way that plants, animals, protists, fungi, and monerans are organized. Your study of ecology and all of the cycles that are involved, makes you think about the earth in a different and more meaningful way.

Although you cannot know everything about everything, you can be a person who solves problems and makes contributions to discussions and debates. You should be able to enter into discussions about genetic engineering, the use of alcohol, drugs, and tobacco, and about the behavior of humans and other organisms. Food webs and food chains are news items in the daily press. You know that oil spills affect organisms, and that lakes, rivers, and bays can become polluted. You know that you can help do something about all these things. You can act as an individual as well as a member of an action group.

CHECK YOUR UNDERSTANDING

1. What would cause the Leakeys to call their fossil find *Homo habilis* and not *Homo sapiens?*
2. Give a good reason for not inventing a new naming system for animals and plants from the fossil record.
3. Describe the relationship between Neanderthals and Cro-Magnons.
4. What is one of the biggest differences between humans and all other organisms?
5. What characteristics did Cro-Magnons share with modern humans?

Summary of Chapter 14

Changes that have taken place in organisms over long periods of time come to life as you "read" the fossil record. The record in the rocks, ice, and minerals shows that living things have been around a very, very long time. The record also shows that even though millions of years have gone by, there is a connection between living things in general. The forms and structure of today's animals bear striking resemblances to animals of the near and distant past.

Evolution is a process that involves change. The changes in animals and plants that gave rise to new species is described in the fossil evidence. Layers of sedimentary rock that are extremely old have fewer kinds of organisms. As the layers get younger and younger, there is evidence of a great spread in the kinds and in the numbers of organisms. The great variety of life on earth today is the product of much change in form and structure. The genes of populations have been passed on for generation after generation.

There are many similarities among all kinds of living organisms. The DNA or genetic material itself is made up of the same kinds of chemicals in all creatures. Changes in DNA and in the proteins that they govern, as well as the built-in clocks of radioactive material, give scientists accurate ways of measuring and studying changes that are taking place and that have taken place in the past.

The theory of evolution offers some explanations for how things have changed. A main idea in the theory is that a process of natural selection is at work. Species produce large numbers of offspring. This provides a chance for genetic variation to take place. The explanations centered around the idea that species are changeable.

The changes are brought on by the variations of genetic material adapting to environmental factors such as radiation. Changes also are built into the population gene pool as the genes themselves cross over, break off, and otherwise undergo internal changes.

Homo sapiens has been one of the most successful of all organisms. The human brain, opposable thumbs, and upright posture have been factors in the success of the "thinking animal." The use of language, the development of culture, and excellent tool-making abilities have put humans at the front of the line.

Humans have made more changes to the earth and have done more to improve their standard of living than any other living thing. Humans have also done much to disturb the balance of life on the planet. Humans have used the earth's resources with a lack of care and concern. No other organism is so much in charge of its own future and destiny. The problem solving, reasoning, and caring organism known as *Homo sapiens* is well equipped to meet the future with much hope and anticipation.

1. *fossils*
2. *paleontologists*
3. *anthropologists*
4. *fossil fuels*
5. *amber*
6. *mineralized fossil*
7. *petrified wood*
8. *adaptations*
9. *geological timetable*
10. *radioactive*
11. *half-life*
12. *evolution*
13. *nucleic acids*
14. *geology*
15. *natural selection*
16. *egg tooth*
17. *Darwin's finches*
18. *Galapagos Islands*
19. *Cro-Magnon*
20. *Neanderthal*

REVIEW QUESTIONS FOR CHAPTER 14

1. Where on the planet have fossils been discovered?

2. What is a mineralized fossil?

3. Describe petrified wood.

4. How is fossil evidence used to support evolution?

5. What can you say about the variety of life forms in the fossil record from 500 million years ago?

6. How does a radioactive material like uranium help give information about the age of fossils?

7. When did the saber-toothed tigers show up in the fossil record?

8. What do the horseshoe crab, the opossum, and the coelacanth fish have in common?

9. What does the fossil record tell about the numbers of species that have lived on the earth?

10. Describe the connection between populations of organisms and the main ideas of evolution.

11. Tell what evolution is about.

12. How does what happens on an island help in the formation of new species?

13. What happens to DNA over long periods of time?

14. List some of the things that Charles Darwin did to arrive at his ideas on the theory of evolution.

15. What is an egg tooth? How does the egg tooth fit with the idea of natural selection?

16. Describe the connection between the story of the peppered moth and the idea of natural selection.

17. What was the relationship between species on the mainland of South America and the species on the Galapagos Islands?

18. Give some reasons for the different species of Darwin's finches.

19. List traits that separate humans from all of the other organisms in the world.

20. Why is tool-making such a special characteristic?

Glossary

abdomen — body cavity in some animals

absorption — taking in materials

adapt — change

adaptation — change in response to the environment

adolescence — period of increased development in humans during puberty

age of reptiles — age of the development of vertebrate animals

algae — plant-like protists that provide food and oxygen for much of the world

alimentary canal — digestive system of vertebrates

allele — opposite form of a gene (tall or short in plants)

alveoli — microscopic air bags in the lungs

amber — yellowish, transparent mineral, sometimes enclosing fossils

ameba — one cell protist that moves by flowing along

amphibian — invertebrate that lives at first in water and later on land

angiosperm — seed plant that produces flowers

animal — living thing that gets its food from other organisms and moves from place to place

animal kingdom — the highest level of the classification system in the group of animals

anther — male organ of reproduction in plants

anthropologist — scientist who digs for and studies fossils

antibodies — proteins in the plasma that fight disease

anus — body opening where wastes pass out

aorta — largest artery that takes oxygen-rich blood from the heart to the body

aquatic — living in water

artery — blood vessel that carries oxygen rich blood away from the heart

arthropod — invertebrate with jointed legs and an exoskeleton

asexual reproduction — the process by which one-celled organisms divide in half to produce two identical daughter cells

association neuron — nerve cell that carries impulses from sensory neurons to motor neurons

atmosphere — layer of gases surrounding the earth

atom — the smallest part of a molecule

auditory nerve — group of sensory neurons that sends information from the ear to the brain

backbone — set of bones called vertebrae

bacteria — monerans that are both helpful and harmful to humans

ball and socket joint — shoulder or hip joint

basic chemical — substance that living things need to stay alive

basic life activity — activity that all living things carry out: getting food, movement, growing, sensing and reacting, using food and removing wastes, development, and reproduction

behavior — things that organisms do

biceps muscle — muscle in the upper arm

bile — chemical that breaks apart fat molecules

biology — science of living things or organisms

biome — large region of the earth containing groups of ecosystems

biosphere — all the biomes on earth

bladder — animal organ that stores urine

blade — flat part of a leaf where food is made

blight — fungus that ruined the potato crop in Ireland in the 1800's

blood platelet — substance in plasma that helps clot blood

blood pressure — measure of how hard blood is pushing against artery walls

blood type — one of four blood types (A, B, AB, or O), caused by different proteins on the red blood cell

blood typing — method of identifying which blood type a person is

blood vessel — artery, vein, or capillary

body circulation — blood flow through the body

bog — wet, spongy ground formed from decayed moss and other plant matter

book lung — spider lung ressembling pages of a book

botanist — scientist who studies and classifies plants

botany — science of plants

bracket fungus — fungus that grows on the sides of trees

brain — control center of the body

brain stem — stem connecting the brain and spinal cord

bronchi — tubes that connect the trachea and lungs

bronchial tube — small tube in the lung

bronchiole — very small tube in the lung

brown algae — giant kelp used to make food and iodine

budding — reproducing as a yeast does by forming a bud which then breaks off to make a new organism

capillary — tiny blood vessel that connects arteries and veins to the cells

carbohydrate — "fuel chemical" that supplies the body with energy

carbon — abundant chemical element of which most things are made

cardiac muscle — involuntary muscle found in the heart

carrier — person who passes a hidden disease to someone else

cartilage — tough tissue which is not as hard as bone

cell — basic unit of life

cell function — jobs that a cell does such as covering or transportation

cell respiration — providing energy to the cells by using oxygen to break down food

central nervous system — brain and spinal cord

cerebellum — brain part that controls balance and coordinates muscular activity

cerebrum — largest brain part that controls thinking, feelings, and body movement

chemical digestion — changing food to a simpler form by mixing it with chemicals

chemical energy — energy from chemicals such as the simple sugars made during photosynthesis

chemical reaction — the breaking up and combining of molecules

chlorophyll — green substance in plants which is used in photosynthesis

chloroplast — plant structure that contains chlorophyl

chordate — having a backbone

chromosome — part in each cell that contains DNA

cilia — hair-like structures that help a paramecium move

circulatory system — the heart, blood vessels and blood moving food and other materials through the body

class — third highest level of the classification system

classification — grouping things into categories according to similar structures

climax community — the last stage in the succession of a community

cloaca — digestive structure in a frog

cochlea — hearing part of the ear

color blindness — sex-linked condition in which people cannot see certain colors

communication — activity of an organism that uses some form of signal or stimulus

community — group of populations of different species living in one place

condense — change from a gas to a liquid

conditioned response — learned behavior

conifer — gymnosperm that produces cones such as pines, firs, and other evergreen trees

consumer — organism that eats another organism

contaminated — made unclean or full of germs

coordination — acting together

cornea — clear, transparent eye covering

courtship behavior — set of patterns used by some animals to attract and acquire a mate

Cro-Magnon — ancestor of *Homo sapiens* that lived about 50,000 years ago

crop — storage place in an insect's digestive system

cross pollination — pollination between two plants

cultural evolution — series of changes in human culture

cycle — series of events that repeat themselves regularly in the same order

Darwin's finches — birds on the Galapagos Islands that led to Darwin's theory that adaptations over long periods of time can lead to genetic changes

daughter cell — new cell that is exactly the same as the original one after asexual reproduction

decompose — break down materials so that they can be used by organisms

decomposer — bacterium, fungus, or other organism that breaks down materials

defense system — body's ability to make antibodies

dehydrate — dry out

denitrifier — bacterium in the soil that breaks down materials in the soil and returns nitrogen to the air

dependent — relying on another for support

development — going through a natural period of growth

diabetes — disease in which the body cannot make insulin, causing too much sugar in the blood

diaphragm — strong muscle under the lungs that helps in breathing

diatom — microscopic alga whose shell is made of silicon

dicot — angiosperm with two embryo leaves such as beans and trees

differentiation — process of cells becoming unlike

diffusion — movement of materials from a crowded area to a less crowded area

digestion — process of changing food to a form that cells can use

disease — sickness; condition in which a body part does not work

dissolve — break apart into small and invisible pieces

diversity — variety

DNA — material which carries a genetic message and controls heredity

dominant (in genes) — visible

dominant behavior — behavior that enables an organism to establish itself as a leader

Drosophila — red-eyed fruit fly used for many genetics experiments

ear — sense organ which picks up sounds and sends information to the brain

eardrum — thin tissue separating the outer and inner ear

ecology — study of ecosystems

ecosystem — interaction of communities of organisms with their non-living environment

egg cell — female cell involved in reproduction

egg tooth — tooth developed by young birds to help break the eggshell

embryo — organism in its early stage of development

endocrine system — system of glands that release chemicals into the bloodstream

energy — ability to produce motion or cause change

environment — an organism's surroundings

enzyme — chemical in an organism that does a special job such as digesting food

epidemic — outbreak of disease affecting many people at the same time

esophagus — tube connecting the mouth and crop in an insect and the mouth and stomach in a human

euglena — one-celled protist that is both like a plant and animal

evaporate — change from a liquid to a gas

evergreen — tree with needlelike leaves that does not lose all its leaves at the same time

evolution — genetic changes in organisms over time

excretion — getting rid of waste materials

excretory system — body system that gets rid of wastes

exhale — breathe out

external fertilization — fertilization outside the body

extinct — no longer existing

eye — complex sense organ that focuses light

fat — "fuel chemical" that provides the body with energy

feathers—the outer covering of birds; one of the "similar structures" that puts animals on the same level of the classification system

fermentation — breaking down carbohydrates by yeast or bacteria

fern — vascular plant with spores instead of seeds

fertilization — the meeting of the pollen tube and the egg in plants

fertilize — mix sperm and egg

fetus — embryo that is at least eight weeks old

first-order consumer — plant eater

fish — animal with scales and a two-chambered heart that lives in water

flatworm — invertebrate that often lives as a parasite inside an animal host

flower — special plant part that makes both egg and sperm cells

food chain — food pattern in which large organisms feed upon smaller or weaker organisms

food molecule — the chemical that food is changed into after digestion

food web — complex pattern of overlapping food chains

fossil — trace of animals and plants of the past

fossil fuel — fuel formed from once living things such as oil or coal

fraternal twin — twin formed when two embryos are fertilized at the same time

fruit — what the ovary becomes after plant fertilization

fungus — (pl. fungi) organism that breaks down and absorbs food from other living or dead things

fur — the outer covering of mammals; one of the "similar structures" that puts mammals on the same level of the classification system

Galapagos Islands — islands off the coast of Ecuador where Darwin conducted research

gall bladder — sac that stores and releases bile

gamete — sex cell

gene — the part of a chromosome that contains DNA

gene pool — all the genes in a certain population

generation — single step in the line from one ancestor to another so that grandparent, parent, and child form three generations

genetic disease — inherited disease

genetic engineering — producing organisms with specific genes

genetics — study of heredity

geneticist — scientist who studies genetics

genotype — the genes of an organism

genus — the level of the classification system that gives each living thing its first part of its scientific name (Ex: *Homo* in the name *Homo sapiens*)

geology — science of the earth

geotropism — response of plants to light

germination — beginning of development

gestation — the period the embryo spends inside the uterus

ginkgo — tree in the gymnosperm group that has fan-shaped leaves and is resistant to disease

gizzard — insect organ that grinds up food

gland — group of cells that make a juice that helps the body work

gonad — sex cell

gosling — newly hatched goose

guard cell — special cell that opens and closes the stomate to let water and gases in and out

gullet — opening by which a paramecium takes in food

gymnosperm — seed plant that produces cones

habitat — home

half-life — amount of time that it takes for one half of a radioactive material to decay

heart — powerful pump that forces blood through the body

heartbeat — rhythm of the heart as it contracts and relaxes

hemisphere — one half of the cerebrum

hemoglobin — substance in red blood cells that attracts oxygen

hemophilia — disease in which blood cannot clot

heredity — passing of traits from parents to children

hinge joint — elbow or knee joint

Homo habilis — fossil human with a smaller brain size than *Homo sapiens*

hormone — chemical messenger that helps regulate body functions

host — organism that a parasite lives on

human culture — all of the knowledge and history related to humans

hybrid — mixed; plants or animals having more than one kind of gene for the same trait

hydrogen — element in water that is split off in the process of photosynthesis

identical twin — twin formed from a single embryo

inbreeding — marrying within a small gene pool

inhale — breathe in

interact — act upon or affect each other

internal fertilization — fertilization inside the body

immunity — ability to resist disease

immunization — injection to cause resistance or immunity to a disease

imprinting — attachment of newborn animals to the first object they see

impulse — small, rapid electrochemical charge that carries information from one neuron to another

infectious — caused by a germ

ingestion — taking in of food or other material

innate — present at birth

innoculate — vaccinate

insulin — hormone in the pancreas that regulates the amount of sugar in the blood

instinct — inborn behavior

invertebrate — animal without a backbone

involuntary muscle — muscle that works automatically

iris — eye muscle that controls light that enters the eye

isolation — being alone

jellyfish — invertebrate with tentacles that mostly floats about in the sea

joint — place where bones meet

kidney — excretory organ that removes wastes from blood

kingdoms — the five main groups into which all living things are divided: animals, plants, protists, fungi, and monerans

large intestine — organ that removes excess water from undigested food and forms solid wastes

learned behavior — behavior that is acquired

left atrium — upper left heart chamber that gets oxygen-rich blood from the lungs

left ventricle — lower left heart chamber that sends oxygen-rich blood through the aorta and to the body

lens — eye part that changes shape to focus light

lichen — partnership of a fungus and an alga living together, with the fungus providing water and the alga providing chlorophyll

ligament — special tissue connecting bones at a movable joint

light energy — energy from the sun

Linnaeus, Carolus — inventor of the classification system for plants and animals

liver — organ that makes bile

living thing — organism that must breathe, react, get food, and carry out the basic life activities

logical patterns — methods of reasoning

lung — organ that connects to the outside air and the inside walls of blood vessels

lung circulation — circulation to and from the lungs

mammal — vertebrate with fur that feeds its young milk from a mammary gland

mammary gland — gland that produces milk after the birth of an offspring; breast

marrow — substance in bones that makes blood cells

marsupial — animal with a very short gestation period that completes its development outside the mother's body

mechanical digestion — changing the size of food particles by chewing

medulla — brain part that controls automatic body activities such as breathing and circulation

meiosis — type of cell division ending with half the number of chromosomes

membrane — outer covering of a cell

Mendel, Gregor — scientist who made important discoveries about heredity

menstruation — discharging of blood from the uterus through the vagina

microbe — microscopic germ

microorganism — very tiny organism that can only be seen under the microscope

mildew — fungus that grows on leather, paper, or cloth

mineral — nutrient necessary for living things

mineralized fossil — fossil enclosed in a mineral

mitosis — process in which a nucleus divides in half

mold — fungus that grows on living and non-living things

molecule — the smallest combination of atoms which form a separate substance

mollusk — marine invertebrate such as the clam and snail with a soft body and shells

moneran — simple organism that has an unusual type of cell structure

monocot — angiosperm such as corn or wheat, with only one embryo leaf

moss — non-vascular plant that grows in damp places on the ground and on the side of rocks and trees

motor nerve or neuron — nerve cell that sends messages from the brain to the muscles

mouthpart — structure that helps an animal eat food

muscle system — system of muscles that moves the body by contracting and relaxing to pull on the bones

mushroom — fungi that grows an umbrella-like cap to reproduce with spores

mutation — change in the genes that can cause abnormalities

mutualism — partnership in which two organisms depend upon each other

natural selection — process of better adapted organisms being able to survive, producing more offspring, and eventually changing a species

Neanderthal — ancestor of *Homo sapiens* that lived about 100,000 years ago

needle — leaf of a conifer

nerve net — a low-level nervous system

nervous system — system composed of the brain, spinal cord, and nerves that controls and coordinates body activities and responds to stimuli

neuron — nerve cell

nitrifier — bacterium in the soil that takes nitrogen from the air and changes it to a useful form

nitrogen — element in the air that bacteria change into useful chemicals

non-living thing — substance that does not carry out the basic life activities

non-vascular plant — plant without transport tubes

nourish — feed

nucleus — control center of a cell

nutrients — the six chemicals — water, carbohydrates, fats, proteins, minerals, and vitamins-that organisms need to stay alive

nutrition — study of foods and how the body uses them

offspring — child or children

optic lobe — insect brain part that coordinates eye movements

optic nerve — group of sensory neurons that sends information from the eye to the brain

organ — basic working part of an organism formed from different types of tissues acting together

organism — living thing

ossification — changing cartilage to bone

ovary — female organ that produces and encloses egg cells

ovulation — production of eggs by the ovaries, usually occurring once each month in humans

oxygen — gas that is necessary for living things

oxygen-carbon dioxide cycle — the continuous process of animals breathing in oxygen and breathing out carbon dioxide and plants taking in carbon dioxide and giving off oxygen

paleontologist — scientist who digs for and studies fossils

pancreas — large digestive gland

parallel — facing the same direction and keeping the same distance apart

paramecium — one-celled, animal-like protist that lives in water and moves by using its cilia

parasite — organism that lives on another organism (the host) and causes harm to it

pasteurized — heated to destroy bacteria

pathogen — germ

Pavlov, Ivan — scientist who used dogs to study behavior

penicillin — medicine from the *Penicillium* mold that kills germs

penis — male organ of reproduction

peripheral nervous system — nerves that send messages between the central nervous system and other body parts

perspiration — waste material that passes through the skin

petiole — leaf part that attaches the leaf to the stem or a branch

petrified wood — wood that has become stone through chemical change

phenotype — what an organism looks like

phloem tube — tube that transports food from the leaves to all parts of a plant

photosynthesis — the process of plants taking in energy from sunlight to make food

phototropism — the reaction of plants leaning toward light

pistil — female organ of reproduction in a plant

placenta — membrane surrounding the fetus where food and wastes are exchanged with the mother

plankton — green ocean algae and small animal life that provide food for many organisms

plant — organism that makes its own food and does not move from one place to another

plasma — fluid or liquid part of the blood

pollen grain — structure in plants housing sperm cells

pollination — the first step in plant reproduction in which pollen is transferred from the stamen to a sticky substance at the top of the pistil

pollutant — material that damages the environment

pollution — undesirable change in an environment

population — individuals of one species in one area

pore — tiny opening in the skin

predator — animal that kills another animal for food

prey — animal taken by other animals for food

producer — organism that makes food; green plant

problem solving — highest kind of learned behavior; finding answers to problems

property — characteristic that tells what something is like

prostate gland — gland that produces part of semen

protein — chemical that builds and repairs body parts and controls body activities

protist — mostly microscopic organism that acts like both plant and animal

protococcus — green alga that lives on land in damp places such as the shady sides of trees

protozoa — animal-like protist that makes its own food and moves about freely

puberty — period in life in which children develop into adults

pubic — relating to the region where the sex organs are located

pulse — regular beating in the arteries caused by contractions of the heart

pupil — colored part of the eye

pure — not mixed

radioactive — giving off rays

reacting — responding to signals from the environment, such as plants turning toward light

reasoning — making connections between ideas

receptor — skin cell that picks up information about heat, cold, pressure, and pain and relays it to the brain

recessive (in genes) — hidden

rectum — last part of the large intestine

recycle — use over again

red blood cell — part of blood that carries oxygen

reduction division — process during meiosis which causes sex cells to divide in half so that each contains half of the species' number of chromosomes

reflex — automatic act

reflex action — automatic response such as blinking

replicate — duplicate

reproduce — produce new organisms

reproduction — the producing of offspring

reptile — egg-laying vertebrate that breathes with lungs

resistant — going against or opposing

respiration — exchange of oxygen and carbon dioxide between cells and the environment

respiratory system — body system that exchanges oxygen and carbon dioxide

response — reaction to a stimulus

retina — part of the eye that receives information from the light rays focused upon it

right atrium — upper right heart chamber that receives oxygen-poor blood from the body

right ventricle — lower right heart chamber that sends blood to the lungs for oxygen

ringworm — fungus infection of the human scalp and skin

root — plant part that supports the plant and absorbs, transports, and stores water and minerals

root cap — structure at the end of each root that protects the root

root hair — thin tube extending from a plant root that absorbs minerals and water from the soil

roundworm — invertebrate such as the hookworm that lives in soil and water

rust — reddish-brown, parasitic fungus that damages the plant on which it lives and turns it to a rusty color

saliva — chemical made by the salivary glands that changes food into a simpler form

salivary gland — group of cells in the mouth that makes a digestive juice called saliva

scales — the outer covering of a fish; one of the "similar structures" that puts fish on the same level of the classification system

sclera — the white of the eye

scrotum — pouch which contains the testes

sea lettuce — marine alga that looks like lettuce

second-order consumer — animal eater

Secretariat — famous race horse

secrete — give off, such as glands giving off chemicals

seed — specialized reproductive part found inside seed plants

segmented worm — invertebrate such as the earthworm that brings air to the soil by tunneling

selective breeding — selecting useful mutations to breed organisms

self pollination — plant pollination within one flower

semen — fluid produced by the prostate gland which mixes with sperm in the urethra

sense organ — body part that gets signals from the outside world

sensing — picking up signals from the environment, such as hearing a fire engine

sensory nerve or neuron — nerve or nerve cell that connects sensory organs to the brain

sessile — attached to one spot

sex chromosome — chromosome that determines the sex of an individual

sex-linked trait — characteristic that usually occurs in only one sex

sexual reproduction — process of producing offspring from two parents, one parent providing the egg cell and the other parent providing the sperm cell

sickle-cell anemia — genetic disorder which affects red blood cells

sign language — method of communication using movement

silicon — glasslike material that a silicon shell is made of

similar structures — parts of organisms that are alike

simple sugars — also called glucose; the food made when plants combine hydrogen and carbon dioxide in the process of photosynthesis

skeletal muscle — voluntary muscle that causes movement

skeletal system — network of bones that supports the body and gives it shape

skin — largest organ and outer covering of the body

Skinner, B. F. — scientist who used rats to study behavior

small intestine — digestive organ in humans in which most digestion takes place

smell — sense that identifies odors by reacting to chemicals

smooth muscle — involuntary muscle found in internal organs and vessels

smut — foul-smelling, parasitic fungus that mostly lives on and damages cereal plants

species — the lowest level of the classification system that gives each organism the second part of its scientific name (Ex. *sapiens* in *Homo sapiens*)

sperm cell — male cell involved in reproduction

spinal cord — the major nerve cord running down the back of vertebrates

spinal nerve — nerve from spinal cord that sends impulses to the body

spinneret — organ in spiders that makes silk threads

spirogyra — a green alga of ponds

spore — reproductive body found in ferns

sponge — simplest of invertebrates, attached to one spot in the water in plantlike fashion

spontaneous generation — theory that life comes from non-living things

sporozoan — animal-like protist; parasite that causes malaria

stamen — male organ of reproduction in a plant

starfish — marine invertebrate whose tubelike feet act as suction cups

stem — plant part that supports leaves, connects leaves and roots, stores food, and transports materials

stimulus — (pl. stimuli) something that activates an organism or part of an organism

stomach — digestive organ between the esophagus and small intestine

stomate — microscopic opening on each leaf that allows gases to enter and leave the leaf; pore

sunlight — energy from the sun; the primary source of energy for the earth

sweat gland — coiled tube that opens at the outside of the skin

synapse — gap or space between neurons

taste — sense that identifies sweet, sour, bitter, and salty

taste bud — receptor on the tongue that reacts to food chemicals

tentacle — long flexible projection that can be used like an arm

territorial behavior — behavior used to claim and defend a territory

testes — organ in males that produces sperm cells

tissue — group of similar cells that act together to do certain jobs

trachea — tube connecting mouth and lungs; windpipe

trait — characteristic

transport tube — conducting tube of vascular tissue that connects plant parts and allows transportation of materials

tripanosoma — animal-like protist; a parasite that lives on the blood of an animal and causes sleeping sickness

tympanic membrane — low-level ear

umbilical cord — cord connecting the embryo with the mother

urethra — tube that takes urine out of the body; tube in the penis where sperm cells travel

urinary bladder — bag that stores urine

urine — waste product formed in the kidneys

uterus — female body part that holds and nurtures a developing embryo; womb

vaccination — injection of weakened or dead microbes causing the body to produce antibodies to a disease

vacuole — bubblelike space in which a protist digests food

vagina — reproductive organ of the female

vascular plant — plant with a system of tubes for transportation of food, water, and minerals

vein (animal) — blood vessel that carries waste products from cells back to the heart

vein (plant) — skinny tube in a leaf, running from the stalk to the stem, that carries food and water

vertebrate — animal with a backbone or set of vertebrae

vitamin — substance necessary for the body to function

voluntary muscle — muscle that you can control

wading bird — bird that walks in water such as the shoebill stork

water — the most plentiful and important chemical in the cells of all organisms, responsible for dissolving and transporting materials

water cycle — constant reusing of water through evaporation and condensation

white blood cell — disease-fighting part of the blood

womb — uterus

xylem tube — tube that transports water and minerals upward in plants

yeast — fungus that reproduces by budding and is used in making bread and wine

zygote cell — new organism formed when a sperm cell fertilizes an egg cell

Index

A

Abdomen 36
Absorb 21, 23
Absorption 70, 74
Adaptation 149, 245, 256
Adolescence 168
Algae 20, 53–56, 59, 69, 73–75, 78
Alleles 196
Amber 256
Ameba 54, 57–61, 65, 194
Amphibians 31, 80, 83, 87
Ancient wing 258
Angiosperm 43
Angiosperms 44, 50
Animal kingdom 28, 30, 32, 39
Animals 14–15, 17–23, 54, 55, 57–59, 62–63, 65, 69–71, 73
Ant lion 79
Anther 189
Anthropologists 255
Antibodies 177–178
Antigens 178
Anus 83–84, 96
Arthropods 35–39
Asexual reproduction 152
Asexually 62–63, 71, 73–74
Automatic action 92

B

Backbone 30–31
Bacteria 63–65, 68, 73, 75
Basic chemicals 3
Basic life activities 2–3, 8–13, 15–19, 21–23, 81
Basic life chemicals 6
Basic units 3
Bees 240, 247, 251
Beetle 79, 81

Behavior 235–241, 248–251
Biologists 196, 204, 206
Biology 15, 23
Biomes 216–217
Biosphere 217, 222, 226
Birds 25–27, 31, 39
Birth 159–160, 162, 166–167, 171
Biston betularia 267
Bladder 84, 96
Blight fungus 68
Blood vessels 85, 87–88, 92, 96
Bog 49, 51
Bone tissue 5
Book lungs 86, 96–97
Botanists 42
Botany 42
Brain 91–94
Brown algae 56–74
Budding 66, 74
Bug 80–81, 94

C

Calcium 181-182
Carbohydrates 7–8, 12–13, 180
Carbon 14 258
Carbon Cycle 224
Carbon dioxide 65
Carrier 198, 202–203
Caterpillar 158
Cell differentiation 157
Cell division 194
Cell functions 4, 13
Cell material 61– 62
Cell membrane 57, 59, 61
Cell respiration 82, 84, 85, 87, 88, 96
Cells 3–9, 12–13, 61–63, 67, 71, 74, 77, 82, 84–85, 87–88, 90, 92
Centipedes 35, 36, 37

Chemical substances 264
Chimpanzee 245
Chlorophyll 53, 55, 58–59, 65–66, 69–70, 73–74
Chloroplasts 55, 58, 59, 60, 65
Chromosome duplication 151
Chromosome theory 196
Chromosomes 149, 151–155, 193
Cicada 80
Cilia 58, 60, 74, 75
Circulation 61, 65, 85–89, 94, 96–97
Classes 31–32, 35, 37, 39
Classification 25, 27, 29–31, 37–38, 41–42
Cloaca 84, 96
Cocoon 158
Coelacanth fish 259
Communication 246–247, 249–251
Communities 212
Conducting tubes 41–42
Conifers 45–46, 50, 51
Consumers 219, 221, 224–225, 227–228
Coordination 61, 89–92, 94, 96
Courtship behavior 236–237
Cro-Magnon 272
Cross-pollination 189
Crustaceans 21
Cultural evolution 249
Cytoplasm 57, 242

D

Darwin, Charles 265–266, 269–270
Darwin's finches 269–270
daughter cell 62
Daughter cells 62–74, 152, 154, 170
Decision-making 249
Decomposers 37 65, 71, 72, 73, 75, 221, 224
Deerfly 80
Dehydrate 180
Development 11, 157, 160–161, 166–168, 170
Diabetes 202, 206–207
Diatoms 57–74, 62, 75
Dicots 44, 46, 50
Differentiation 166
Diffusion 61, 65, 75

Digestion 9, 13, 60, 81–85, 94, 96–97
Diversity 149, 152, 170
DNA 148, 150, 154, 193, 201, 264, 275
Dominant behavior 238
Dominant gene 190
Dragonfly 79
Drosophila 197, 198

E

E. coli 64
Ecology 211–212, 214
Ecosystem 211–212, 214–216, 218, 221, 226, 231
Egg cells 152–154, 171
Egg tooth 267
Embryo 44, 49, 50, 51, 156, 159, 160
Energy 7, 8, 9, 12
Environment 187, 199–200, 202, 211–212, 214, 216–217, 231, 263, 266, 268–269, 275
Enzymes 59–60, 66, 70–71, 74–75, 81–84, 94, 97, 261
Essential needs 219
Euglena 54, 58–59, 60–61, 74–75
Evolution 256, 262, 264–266
Excretion 9, 13, 60, 81–83, 85, 94, 96
Exoskeleton 35
External fertilization 155
Extinct 148, 170
Extinction 263

F

F_1 generation 191
False feet 57, 59
Fats 7, 8, 12, 13
Feathers 26
Fermentation 67, 74, 75
Ferns 41–43, 46–48, 50
Fertilization 154, 160, 162, 165
Flagellum 54, 58, 63
Flowering plants 43–46, 50
Food chain 211, 218, 220–221, 224, 229
Food pyramid 220

Food web 211, 220-221
Food-getting 77–78, 81, 90–91
Fossil fuels 256
Fossils 255–261
Four basic food groups 182–183
Fuel chemicals 7, 13
Fungi 14– 23, 53–54, 65–75
Fur 26, 29

G

Galapagos Islands 269
Gametes 153–154, 194–193
Gene pool 199, 203, 207
Generation 190–191
Genes 193–194, 196, 198–200, 202, 204, 205–
    207, 262–264, 267, 275
Genetic diseases 199, 209
Genetic engineering 206, 207
Genetic material 264
Geneticists 200, 204
Genetics 188, 193, 197, 199, 204, 207
Genotype 191
Genus 28, 29, 31, 32, 38, 39
Geotropism 241
Germs 176, 178
Gestation 160
Ginkgo 46
Ginkgo 46, 51
Gizzard 83, 96
Gonads 152
Goslings 238–239
Green algae 78
Growing 3, 9, 10, 11, 12
Gullet 60, 74
Gymnosperms 45–46, 50

H

Habitat 213
Half-life 258
Hemophilia 203, 207
Heredity 187–188, 193, 199–200, 204, 235,
    237, 251
Homo sapiens 249, 271, 272

Horse 256, 259
Horsefly 80
Horseshoe crab 259
Human being 28, 187, 217
Human culture 249
Human fertilization 165
Human species 271, 273
Humans 235, 255
Humus 49–50
Hydra 33, 78, 82, 85–87, 89–90, 93

I

Imprinting 238, 242, 251
Inbreeding 203, 207
Infectious 175, 178, 183
Ingested 78
Ingesting 77, 94
Innate 244, 251
Innate behavior 236, 251
Insects 79–83, 86, 87, 89, 91, 93–95, 97
Insulin 202, 206
Internal fertilization 155
Invertebrate organisms 256
Invertebrates 32–33, 35, 38–39

J

Jellyfish 33, 78, 85–86, 90, 93, 94

L

Language patterns 246
Larvae 79
Leakey, Mary and Louis 272
Learned behavior 235, 242–246, 251
Lichen 69–70, 72, 74–75
Linnaeus, Carolus 42
Living things 2–19, 22–23, 211, 217, 219, 222,
    224–225, 231
Logical patterns 245
Lorenz, Konrad 239
Lucy 271

M

Mammal embryos 265
Mammals 26, 29, 31, 148, 157, 162, 171
Mammary glands 162
Marsupial 161
Meiosis 153–154, 170, 195
Mendel, Gregor 188, 189–192
Mendel's laws 192
Menstruation 164
Microbes 176, 178
Microorganisms 20, 176
Mildew 70, 74–75
Millipedes 35, 36, 37
Mimosa 241
Mineralized fossil 256
Minerals 180, 183
Mitosis 62, 73–75, 151, 154, 170, 194
Mollusks 34, 35
Monera 54, 73
Monerans 14–22, 53–54, 63, 73
Moss 41, 42, 43, 48, 49, 50
Moss Reproduction 49
Motor nerve 93– 94, 96, 97
Movement 4, 5, 10, 12, 13
Multi-cellular animals 77
Muscle tissue 5
Mushrooms 67, 69, 74
Mutations 200, 202, 204, 207
Mutualism 69, 71, 72, 74

N

Natural selection 266–268, 275
Neanderthal 272
Nerve tissue 5
Nervous system 89–97
Nitrates 225
Nitrifiers 225
Nitrogen 64, 65, 73
Nitrogen cycle 225
Non-flowering plants 45, 50, 51
Non-living things 14, 16–18, 22–23, 212
Non-vascular organisms 65
Non-vascular plants 42–43, 48, 50

Nourish 180
Nucleic acid 264
Nucleus 149, 154
Nutrients 81
Nutrition 180, 183

O

On the Origin of Species 265
Opossum 259
Optic lobe 91, 96–97
Organism 15, 17–18, 23, 56, 58, 59, 64, 71, 75, 152, 157, 211, 212, 213, 156, 217–219 221, 225, 227, 229, 230, 231
Organisms 53–71, 73, 78, 82, 86, 91, 187, 194, 199, 204–205, 207, 260, 262–266, 273– 274
Organs 5, 6, 11, 12, 13
Ovaries 154, 163
Ovary 152, 163
Ovulation 163
Oxygen 78, 85, 86, 87, 88, 89, 94, 97
Oxygen cycle 224

P

Paleontologists 255, 260
Parallel 44, 45, 51
Paramecium 54, 58, 59, 60, 61, 75
Parasites 21, 23, 75
Pathogens 176, 183
Pavlov, Ivan 243, 244
Peat moss 49
Penicillin 68, 70, 74
Penis 162, 165
Phenotype 191
Phosphorus 181–182
Photosynthesis 55, 61, 65
Phototropism 240-241
Pigmy plants 49
Pistil 189
Placenta 159
Planaria 33
Plankton 78

Plants 14–15, 17–23, 53–54, 57, 59, 63–65, 68, 70–71, 73
Pollution 216
Population 212–213, 262, 275
Potassium 182
Predators 78
Pregnancy 165, 167
Prey 78–79, 82–83, 91
Primates 245, 248
Problem-solving 242, 246, 249
Producers 219–221, 224, 227–231
Properties 14, 16–17, 23
Prostate gland 162, 165
Proteins 7, 8, 12, 13, 180, 183
Protist 14–23 53–55, 57–63, 65, 73, 75, 78, 211, 228, 231
Protococcus 56, 74–75
Protozoa 53–54, 57, 59, 73
Puberty 168
Pubic 168
Punnett square 190–191

R

Radioactive 257–258, 264, 275
Reacting 10–12
Reasoning 242, 245–246, 248
Recessive gene 202, 207
Reduction division 194–195, 207
Reflex 236, 243
Regeneration 33
Replicate 201
Reproduction 11, 13, 62, 63–64, 66, 71–72, 148–173, 262, 263
Reptiles 27, 31, 39
Respiration 9, 13, 82, 85–88, 96
Response 236, 240–245, 251
Ringworm 69, 72, 74
RNA 260
Roundworms 33
Rust 68

S

Saliva 243, 244, 245
Salt water 7
Sanitation 178
Scales 26, 27, 39
Scrotum 162
Sea lettuce 56, 75
Second-order consumers 219, 228–229
Sedimentary rock 259, 264, 274
Seed coats 43
Seed plants 41–43, 47–48, 50
Segmented worms 34
Selective breeding 204, 205, 207
Self-pollination 189
Sensing 3, 9, 10, 11, 12, 13
Sensory nerve 93, 96, 97
Sessile 77, 78, 96
Sex chromosome 196
Sex-Influenced traits 203
Sex-linked traits 198
Sexual organs 168
Sexual reproduction 71, 148, 150, 152, 162, 170, 187, 194, 196
Sickle-cell anemia 202
Sign language 248
Silica 57
Silicon 56
Similar structures 25, 29, 35, 37, 38
Skinner, B.F. 244
Smallpox 175, 177
Smut 68, 74
Sodium 182
Softwood 46, 51
Species 28–39, 42–48, 50
Sperm 154
Sperm cells 152, 165
Sphagnum 49
Spider 81–83, 85–87, 89, 91–95, 239, 251
Spinal cord 31, 92, 94, 96
Spinneret 91, 96
Spirogyra 55, 74–75
Spoilage 64
Sponges 32–33, 39, 77–78, 81–82, 87, 89–90, 93–94, 97

Spontaneous generation 148
Spore 58, 63, 65, 66–68, 70–72
Spores 46, 47, 48, 49, 50, 51
Sporozoan 58, 63, 74
Stages of development 156, 158, 167
Starfish 34, 39
Stimuli 61, 62, 74, 240, 242, 251
Stimulus 247
Strontium 90 258
Symbiosis 69

T

Tentacles 78, 82, 90, 94, 96
Termite 79
Testes 152, 154, 162
Thorax 36
Tissue 5, 6, 11, 12, 13
Traits 188, 192, 193, 263, 263, 266
Transport tubes 42
Trypanosoma 58
Tube-like structures 42
Tympanic membrane 91, 96, 97

U

Umbilical cord 166–167
Uranium 257-258
Urethra 162
Uterus 159, 161–163, 165, 167

V

Vaccination 177
Vacuole 60, 65, 74
Vascular plants 42, 46–47, 50
Venus flytrap 241
Vertebrae 30, 31
Vertebrates 30, 32, 38–39, 83–84, 86–87, 89, 97
Vitamins 180, 183

W

Walking stick 79
Water 6, 7, 8, 9, 12
Water cycle 223, 226
Web 79, 91, 92, 95

Y

Yeast 66, 67

Z

Zygote 154–155